Mustapha Sarra

Electrotechnics: Principles, Circuits, and Applications

Mustapha Sarra

Electrotechnics: Principles, Circuits, and Applications

LAP LAMBERT Academic Publishing

Imprint
Any brand names and product names mentioned in this book are subject to trademark, brand or patent protection and are trademarks or registered trademarks of their respective holders. The use of brand names, product names, common names, trade names, product descriptions etc. even without a particular marking in this work is in no way to be construed to mean that such names may be regarded as unrestricted in respect of trademark and brand protection legislation and could thus be used by anyone.

Cover image: www.ingimage.com

Publisher:
LAP LAMBERT Academic Publishing
is a trademark of
Dodo Books Indian Ocean Ltd. and OmniScriptum S.R.L publishing group

120 High Road, East Finchley, London, N2 9ED, United Kingdom
Str. Armeneasca 28/1, office 1, Chisinau MD-2012, Republic of Moldova, Europe
Managing Directors: Ieva Konstantinova, Victoria Ursu
info@omniscriptum.com

Printed at: see last page
ISBN: 978-620-8-44003-9

Preface

This book, **Electrotechnics: Principles, Circuits, and Applications**, provides a comprehensive introduction to the fundamental concepts of electrical engineering. It is designed to serve as both a structured learning resource and a reference for anyone seeking to develop a solid understanding of electrotechnics, whether for academic study, professional development, or self-learning.

The book is structured into seven key chapters:

- **Chapter I: Mathematical Review of Complex Numbers** – Covers fundamental operations with complex numbers and their applications in electrical engineering.
- **Chapter II: Review of the Fundamental Laws of Electricity** – Introduces essential laws such as Ohm's Law, Kirchhoff's Laws, and their role in circuit analysis.
- **Chapter III: Electric Circuits and Power** – Explores circuit components, network theorems, and power calculations in electrical systems.
- **Chapter IV: Three-Phase Electrical Systems** – Discusses three-phase circuits, their advantages, and methods of analysis.
- **Chapter V: Magnetic Circuits** – Examines magnetic materials, magnetic field concepts, and their applications in electrotechnical systems.
- **Chapter VI: Single-Phase Transformers** – Presents the principles of transformers, equivalent circuits, and performance analysis.
- **Chapter VII: Introduction to Electrical Machines** – Provides an overview of electrical machines, including their classification and basic operating principles.

While this book aims to present essential theoretical and practical knowledge, it does not claim to be exhaustive. The material is based on well-established textbooks and references, as listed in the bibliography. These sources have guided the content development, ensuring a solid foundation in electrotechnics.

Since electrical engineering is a constantly evolving field, this book will be regularly updated with improvements, additional examples, and new exercises. Readers are encouraged to provide feedback, corrections, and suggestions to enhance future editions. Furthermore, consulting additional references and online resources is highly recommended for a deeper understanding of the subject.

By balancing theoretical principles with real-world applications, this book aims to equip readers with the necessary analytical and practical skills to understand, design, and analyze electrical systems. It is intended as a dynamic and evolving learning tool, encouraging further exploration and innovation in the field of electrotechnics.

Table of Contents

Mathematical Background on Complex Numbers

Table of contents

I- Definitions

There exists a set denoted by C, called the set of complex numbers, which has the following properties:

- C contains the set of real numbers *R*
- Addition and multiplication of real numbers are extended to complex numbers, and the calculation rules remain unchanged.
- There exists a complex number denoted i such that: $i^2 = -1$.
- Any complex number z can be written in a unique way $z = x + yi$ with x and y real.
- The writing $z = x + yi$ is termed **algebraic form** or **Cartesian form** of the complex number z. x is the real component of z, abbreviated ***Re(z)***, y is the imaginary part of z labeled ***Im(z)***.

Note: $z = x + yi$ with x and y real numbers

- If y = 0, the complex number is real.
- If x = 0, the complex number is called a pure imaginary number

Property :

Assume that $z = x + yi$ and $z' = x' + y'i$

If $z = z' \Leftrightarrow \begin{cases} x = x' \\ y = y' \end{cases}$

This means that a complex number can be **uniquely** written in algebraic form.

- z is a **pure real number** if and only if ***Im(z)=0.***
- z is a **pure imaginary number** if and only if ***Re(z)=0***

II- Graphical representation of a complex number

$(o, \vec{u}, \vec{v})$ is a direct orthonormal coordinate system of the plane.

- For each complex number $z = x + yi$, where x and y are real, we associate the point M with coordinates (x; y). The image point of z is denoted by M, while the image vector of z is denoted by $\overrightarrow{OM}$.

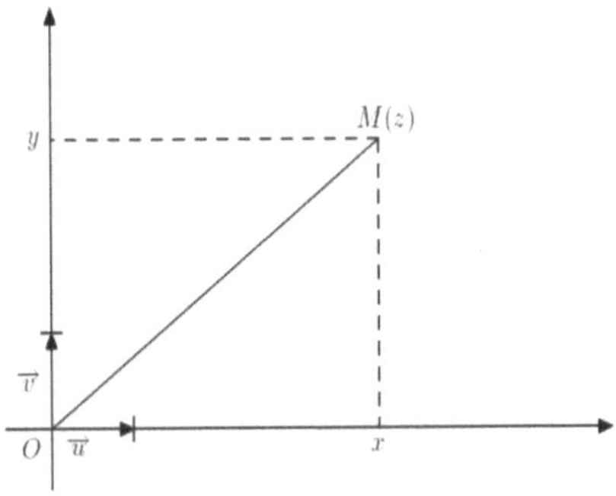

Figure 1.1 Graphical representation of a complex number

- Every point M(x; y) is the image point of a unique complex number $z = x + yi$.
- We define z as the affix of the vector $\overrightarrow{OM}$ and the point M.
- Le plan est alors appelé plan complexe.
- The x- axes $\overrightarrow{ou}$ is called **the real axis,**
- The y-axes $\overrightarrow{ov}$ is called **the pure imaginary axis.**

III- Complex Numbers Operations

III-1 Addition and multiplication

If $z = x + yi$ and $z' = x' + y'i$ are two complex Numbers and if k is a real number ,

$$\text{So : } \begin{cases} z + z' = (x + x') + i(y + y') \\ kz = kx + iky \\ zz' = (xx' - yy') + i(xy' + x'y) \end{cases}$$

Properties

- **Commutativity :** $z + z' = z' + z$ et $zz' = z'z$
- **Associativity :** $(z + z') + z'' = z + (z' + z'')$ et $(zz')z'' = z(z'z'')$
- **Distributivity :** $z(z' + z'') = zz' + zz''$

Notes:

1- The binomial identities (Formulas) are still valid in C. Then We have for all complex numbers z and z': $z^2 + z'^2 = z^2 - (iz')^2 = (z + iz')(z - iz')$.

2- In the complex plane, let M and M' be points with affixes z and z', respectively.

The affix of point P, such that quadrilateral OMPM' forms a parallelogram, is z+z'.

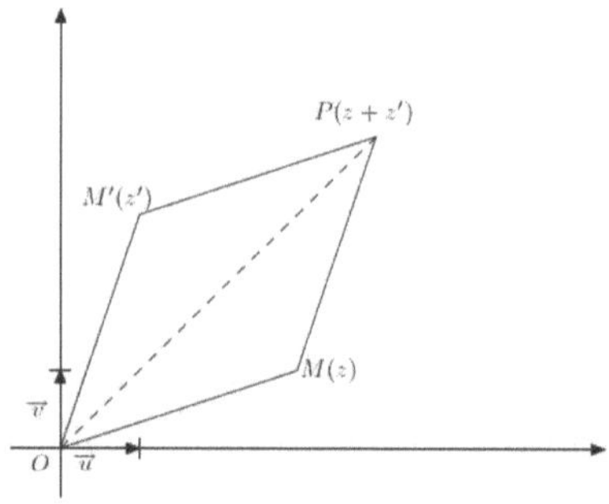

Figure 1.2 Geometric Interpretation of Complex Number Addition

III-2 Inverse of a non-zero complex number

Any non-zero complex number z, expressed in algebraic form $z = x + iy$, has an inverse, denoted $\frac{1}{z}$ equal to $\frac{1}{z} = \frac{x - iy}{x^2 + y^2}$

Dividing by a non-zero complex number z is equivalent to multiplying by its inverse $\frac{1}{z}$.

III-3 Conjugate of a complex number

Let z be a complex number, $z = x + iy$. The conjugate number of z, denoted $\bar{z}$, is the complex number $\bar{z} = x - iy$. In the complex plane, the point M' with affix $\bar{z}$ is the image of the point M with affix z by the symmetry with respect to the abscissa axis.

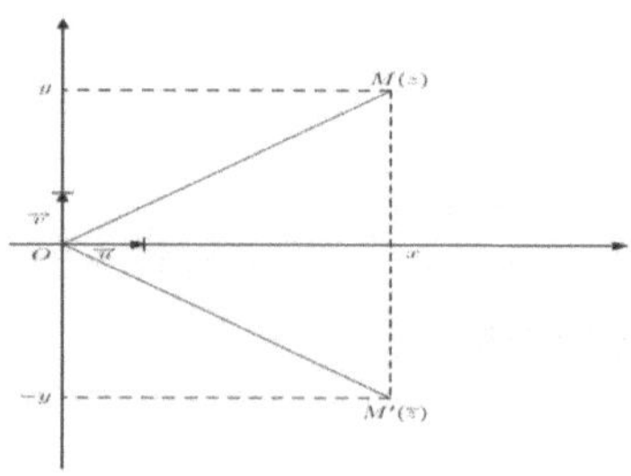

Figure 1.3 Geometric Representation of a Complex Number and Its Conjugate

Note:

z is a complex number.

1- z is real is equivalent to z = $\bar{z}$

2- z is pure imaginairy equivalent to $\bar{z}$ = −z.

IV - Module and argument of a complex number

a- Complex number module

z is complex number, $z = x + iy$ (x and y reals). Module of z is the positive real number denoted $|z|$ and defined by: $|z| = \sqrt{x^2 + y^2}$

Geometric interpretation :

In the complex plane, if M has affix z, then OM = | z |.

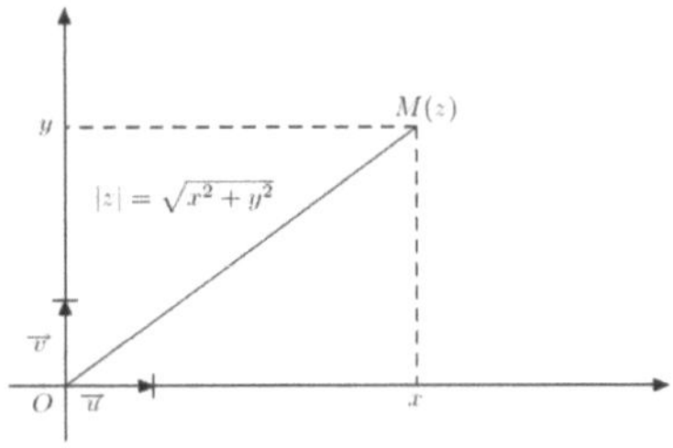

Figure 1.4 Modulus of a Complex Number in the Complex Plane

Notes:

1. If z is a **real number**, The module of z corresponds to the absolute value of z.
2. | z | = 0 equivalent to z = 0 because OM = 0 equivalent to O = M.
3. $z.\bar{z} = x^2 + y^2 = |z|^2$

Examples :

1- Module of Z = -3 + 4i : $|Z|^2 = 9 + 16 = 25$ then | Z | = 5.

2- Module if Z = 9i : | Z | = 9.

3- We give a complex number $Z_A = -1 + 3\mathbf{i}$ and $Z_B = 2 - \mathbf{i}$. *A* is the image of Z_A ; *B* is the image of Z_B ;

Calculate the distance AB.

The affixe of vector $\overrightarrow{AB}$ is $Z_B - Z_A = 3 - 4i$, so $AB = |Z_B - Z_A| = \sqrt{3^2 + (-4)^2} = 5$

b- Complex number argument

Any measure of the oriented angle $(\vec{u}, \overrightarrow{OM})$ is called an argument of a non-zero complex number Z.

It is noted:

$$\theta = arg(Z) = \text{tg}^{-1}\frac{y}{x}$$

Example

The main argument of the complex number $z=-2\sqrt{3}+2i$ is: $\theta = tg^{-1}(\frac{2}{-2\sqrt{3}}) = -30°$

Notes:

1- To multiply two non-zero complex numbers, we multiply the modules and add the arguments.

2- To divide two non-zero complex numbers, we divide the modules and subtract the arguments.

V - Trigonometric and exponential form of a complex number

From Figure 4, we can deduce that for the complex number $z=x+iy$, $\begin{cases} x=|z|\cos\theta \\ y=|z|\sin\theta \end{cases}$

So $\boxed{z=|z|(\cos\theta+i\sin\theta)}$ whith $\theta=\arg(z)$

This writing is called the trigonometric form (or polar form) of the complex number z.

The complex number $z=x+iy=|z|(\cos\theta+i\sin\theta)$ can also be written in the form $z=|z|e^{i\theta}$, it's the **exponential form of z**

Example :

1- Determine the algebraic form of the following complex $z=\frac{(1+i)^4}{(\sqrt{3}-i)^3}$

Solution :

$$z=\frac{(1+i)^4}{(\sqrt{3}-i)^3}=\frac{\left[\sqrt{2}e^{i\frac{\pi}{4}}\right]^4}{\left[2e^{-i\frac{\pi}{6}}\right]^3}=\frac{4e^{i\pi}}{8e^{-i\frac{\pi}{2}}}=\frac{1}{2}e^{i\frac{3\pi}{2}}=\frac{1}{2}e^{-i\frac{\pi}{2}}=\frac{1}{2}(\cos(-\frac{\pi}{2})+\sin(-\frac{\pi}{2}))=-\frac{1}{2}i$$

VI - De Moivre's and Euler's Formulas

De Moivre Formula:

For $\theta \in \mathrm{R}$ and for $\mathrm{n} \in \mathrm{Z}$:

$$\theta \in \mathrm{R} \text{ and } \mathrm{n} \in \mathrm{Z} : (\cos\theta+i\sin\theta)^n=\cos(\mathrm{n}\,\theta)+i\sin(\mathrm{n}\,\theta)$$
$$\text{and } (\cos\theta-i\sin\theta)^n=\cos(\mathrm{n}\,\theta)-i\sin(\mathrm{n}\,\theta)$$

Euler's Formulas :

They are given by : $e^{i\theta} = \cos\theta + i\sin\theta \Rightarrow \cos\theta = \frac{e^{i\theta} + e^{-i\theta}}{2}$ and $\sin\theta = \frac{e^{i\theta} - e^{-i\theta}}{2}$

VII – Applications of complex numbers in electricity

Complex numbers are fundamental in the analysis and design of electrical systems, particularly in alternating current (AC) circuits.

In electricity, we can characterize the behaviour of a linear passive dipole in **sinusoidal mode** with a complex number called **complex impedance**

1- Voltage

The sinusoidal quantity written in the general form:

$$u(t) = U\sqrt{2}\sin(\omega t + \theta_u) \Rightarrow \underline{u} = Ue^{j\theta_u}$$

is associated with the complex number : $\underline{u} = Ue^{j\theta}$ which can also be noted $\underline{u} = (U;\theta)$

As for the Fresnel representation, the modulus is the effective value U and the argument the phase at the origin θ.

2- Current

The same goes for current.

$$i(t) = I\sqrt{2}\sin(\omega t + \theta_i) \Rightarrow \underline{i} = Ie^{j\theta_i}$$

Note:

In electricity, we use ***j*** as a complex variable instead of ***i*** to avoid confusion with the current ***i***.

Generally speaking, if we consider a dipole of impedance z.

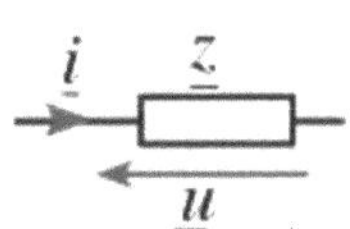

The complex impedance is expressed as: $\underline{z} = Ze^{i\varphi}$

Z is the impedance in ohms (Ω)
φ est le déphasage provoquée par le dipôle entre la tension aux bornes du dipôle et le courant qui le traverse (en radians - rad).

Ce qui donne pour les dipôles R, L et C

	Resistance	Inductance	Capacitance
	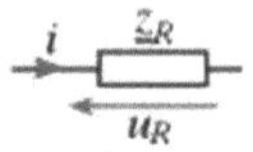	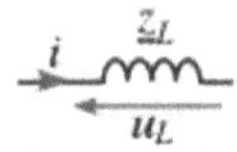	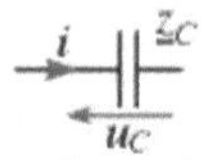
Impedance (Ω)	$Z_R = R$	$Z_L = L\omega$	$Z_C = \frac{1}{C\omega}$
Phase-shift (rad)	$\varphi_R = 0$	$\varphi_L = +\frac{\pi}{2}$	$\varphi_C = -\frac{\pi}{2}$
Complex Impedance (Ω)	$\underline{Z}_R = R$ Real	$\underline{Z}_L = jL\omega$ Pure imaginairy	$\underline{Z}_C = -j\frac{1}{C\omega}$ Pure imaginairy
Admittance complexe (Siemens - S)	$\underline{Y}_R = \frac{1}{R}$ Real	$\underline{Y}_L = -j\frac{1}{L\omega}$ Pure imaginairy	$\underline{Y}_C = jC\omega$ Pure imaginairy

VIII- Representation of AC Signals:

1- AC signals are typically sinusoidal and vary with time. Using complex numbers, these signals can be represented in a simplified and compact form as phasors.

For example: $v(t) = V_m.\cos(\omega t + \phi)$ is represented as $v(t) = V_m e^{j\phi}$

Here:

- V_m : peak amplitude,
- ω : angular frequency,
- ϕ: phase angle.

2. Impedance in AC Circuits

Impedance Z is a complex quantity that combines both resistance (R) and reactance X_C

The total impedance in an AC circuit can be expressed as: $\underline{Z} = R + jX_L - jX_C$

Where:
- R is the real part (resistance),

- $X_L = L\omega$ is inductive reactance, which can be or capacitive).
- $X_C = -\frac{1}{C\omega}$ is capacitive reactance,
- $\omega = 2\pi f$ is the angular frequency,
- L is inductance,
- C is capacitance.

Example:

Consider an AC circuit with a 10 Ω resistor and a 2 H inductor at 50 Hz. The inductive reactance is:

$$X\_L = j\omega L = j(2\pi * 50)(2) = j628.3\ \Omega$$

So, the total impedance is:

$$Z = R + jX\_L = 10 + j628.3\ \Omega$$

IX- Conclusion

This chapter has provided a comprehensive review of complex numbers, covering their definitions, graphical representations, algebraic operations, modulus and argument, as well as their trigonometric and exponential forms. Special emphasis was placed on De Moivre's and Euler's formulas, which are fundamental in various mathematical and engineering applications.

The significance of complex numbers in electrical engineering was also explored, particularly in the analysis of AC circuits. The use of complex impedance simplifies circuit calculations, allowing for an intuitive understanding of voltage, current, and phase relationships in sinusoidal steady-state conditions.

Mastering the mathematical properties of complex numbers is essential for advanced studies in control systems, signal processing, and electrotechnics. These concepts serve as a foundation for analyzing and designing more complex electrical and electronic systems.

Review of the Fundamental Laws of Electricity

Table of contents

I. Introduction

1. Direct Current (DC) Regime (Static Regime):

In this regime, the electrical quantities (currents and voltages) of a circuit are time-independent (constant).

Inductances (coils) are treated as **closed switches**, and **capacitors** are treated as open switches. Consequently, **resistances** are the only linear components considered in the analysis.

The analysis of such circuits relies solely on **Kirchhoff's laws and Ohm's law.**

Dans ce régime, les grandeurs électriques (courants et tensions) d'un circuit sont indépendantes du temps (constantes).

2. Sinusoidal Regime (Harmonic Regime):

An electrical circuit is said to be in a sinusoidal steady-state regime when the external excitations (currents or voltages) are **sinusoidal functions** that induce **currents and voltages of the same form** within the circuit.

3. Transient Regime

Transient phenomena in electrical quantities occur during sudden changes in the characteristics of an electrical circuit.

II. Dipoles in Direct Current (DC) Regime

1- Resistive dipole

A resistive dipole is a passive component. The energy it absorbs is entirely consumed through **Joule heating** and is therefore dissipated **as heat.**

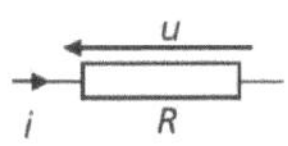

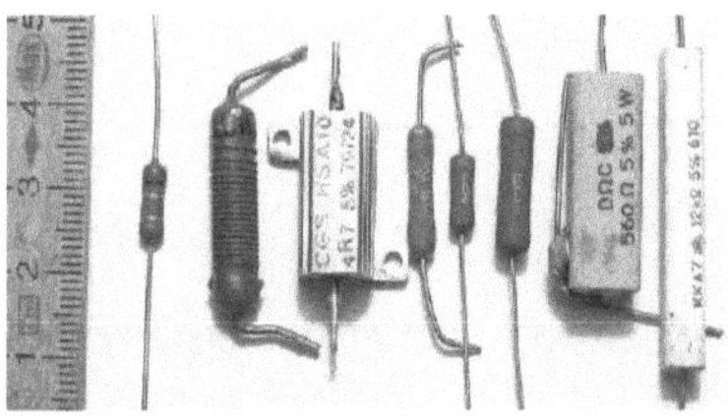

When a resistive dipole is traversed by a current of intensity I, an electric voltage U appears across its terminals. By assigning various values to the current I, it is observed that the voltage U is proportional to the current I.

The proportionality constant is called the resistance of the resistive dipole, denoted as R, and is expressed in ohms (Ω).

Furthermore, the resistance of a conductive wire depends on the material properties of the conductor and its dimensions. It is calculated using the relation:

$$R = \rho \frac{l}{s}$$

R : The value of the conductor's resistance in ohms (Ω),
ρ: The value of the resistivity of the conductor $(\Omega.m)$,
l : Conductor length (m),
s : Conductor Section (m²).

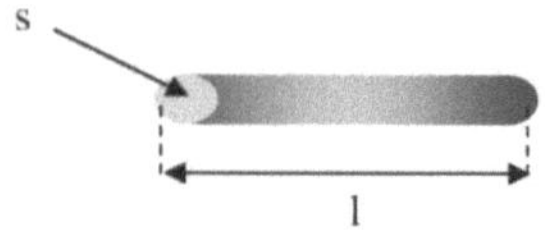

a- Ohm's law

The representation of the electrical voltage U which appears at the terminals of a resistive element crossed by a current of intensity I is given below.

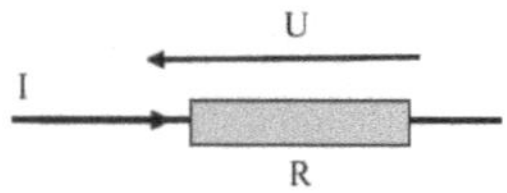

La tension U s'exprime en fonction de l'intensité I du courant par la loi d'Ohm :

$$U = R.I$$

b- Series Combination of Resistive Elements

When multiple resistive dipoles are connected in series, they are traversed by the same current. They can be replaced, to make the calculations more manageable, by a single equivalent resistance, called the equivalent resistance (***R$_{eq}$***).

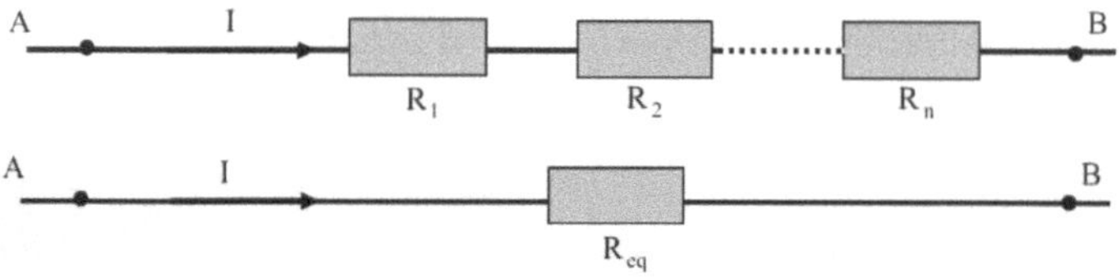

The equivalent resistance is calculated using the following relation: $R_{eq} = \sum_{i=1}^{n} R_i$

Ohm's law gives: : $U = R_{eq}.I$

c- Parallel Combination of Resistive Elements

When multiple resistive dipoles are connected in parallel, they are subjected to the same voltage.

They can be replaced by an equivalent resistance (R_{eq}).

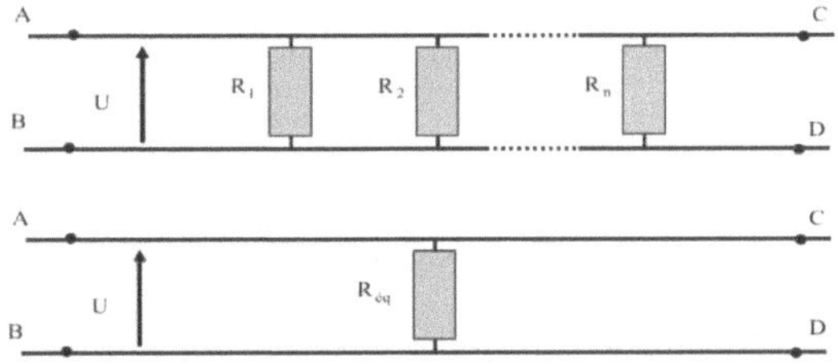

The equivalent resistance is calculated using the following relation

$$\frac{1}{R_{eq}} = \sum_{i=1}^{n} \frac{1}{R_i}$$

2 - Capacitive dipole

The capacitor is a basic electronic or electrical component, consisting of two conductive plates (called "electrodes") separated by a dielectric (insulator). Its main property is its ability to store electric charge. The capacitor is characterized by its capacitance C.

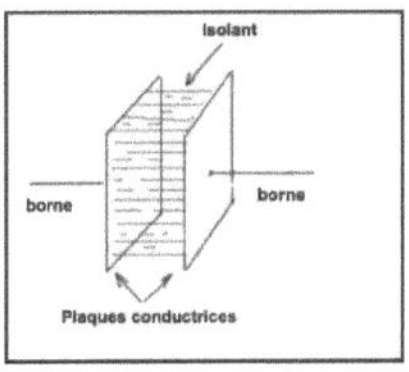

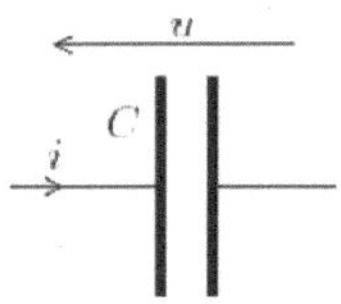

The characteristic relation of an ideal capacitor is: $i_c = C\frac{du_c}{dt} \Rightarrow u_c = \frac{1}{C}\int i_c dt$

Where:

- ***i(t)*** is the current through the capacitor,
- ***u(t)*** is the voltage across the capacitor,
- ***C*** is the capacitance of the capacitor,
- $\frac{du_c}{dt}$ is the derivative of the voltage with respect to time.

This equation describes the relationship between the current and the voltage in an ideal capacitor.

Note:

In the direct current (DC) regime, the capacitor behaves like an open switch, meaning the current flowing through it is consequently zero.

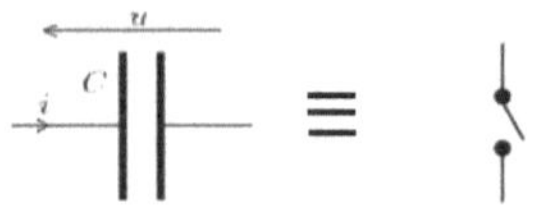

- **Series Combination of Capacitors:**

When capacitors are connected in series, the total or equivalent capacitance C_{eq} is calculated using the following relation: $\frac{1}{C_{eq}} = \frac{1}{C_1} + \frac{1}{C_2} + ... + \frac{1}{C_n}$

Where C_1, C_2 ... C_n are the individual capacitances of the capacitors in series. The total capacitance in a series combination is always smaller than the smallest capacitance in the series.

$$\frac{1}{C_{eq}} = \sum_{i=1}^{n} \frac{1}{C_i}$$

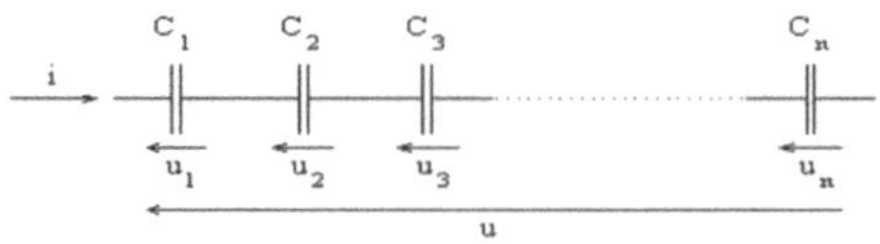

- **Parallel Combination of Capacitors:**

When capacitors are connected in parallel, the total or equivalent capacitance C_{eq} is the sum of the individual capacitances: $C_{eq} = C_1 + C_2 + ... + C_n$

Where C_1, C_2 ... C_n are the individual capacitances of the capacitors in parallel. In this configuration, the total capacitance is always greater than the largest capacitance in the parallel group.

$$C_{eq} = \sum_{i=1}^{n} C_i$$

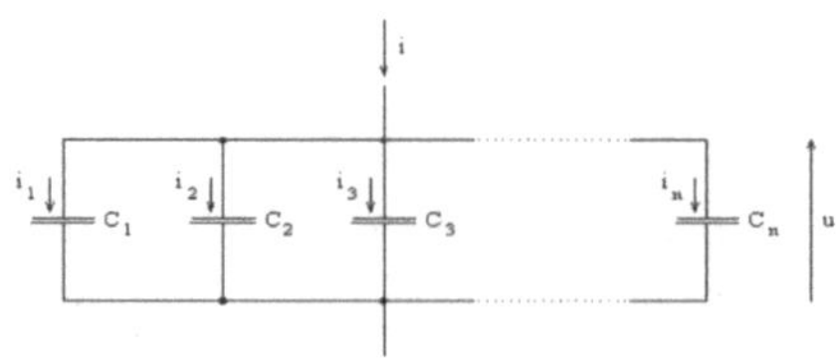

2- Inductive Dipole

An inductor (or coil) is generally made of a conductive wire wound into a helical shape, forming a solenoid.

When an inductor is traversed by a current, it generates an induced magnetic field that opposes the current that created it **(Lenz's Law)**.

The inductor is characterized by its **inductance L**, defined as the ratio of the magnetic flux to the current passing through it.

Inductance L is expressed in henrys (H).

The voltage across the terminals of an inductor is related to the current intensity by the following relation: $u_L = L\frac{di_L}{dt}$

Its schematic representation is as follows:

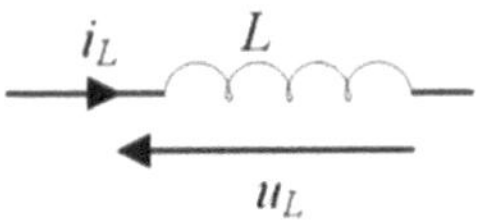

Note :

In a DC steady-state, the coil behaves like a closed switch, and consequently, the voltage across its terminals is zero.

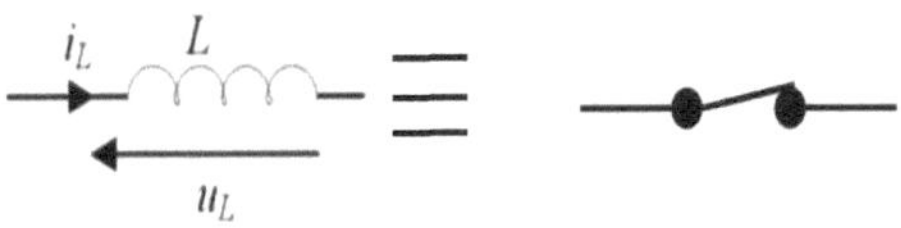

- **Series connection of coils:**

$$L_{eq} = \sum_{i=1}^{n} L_i$$

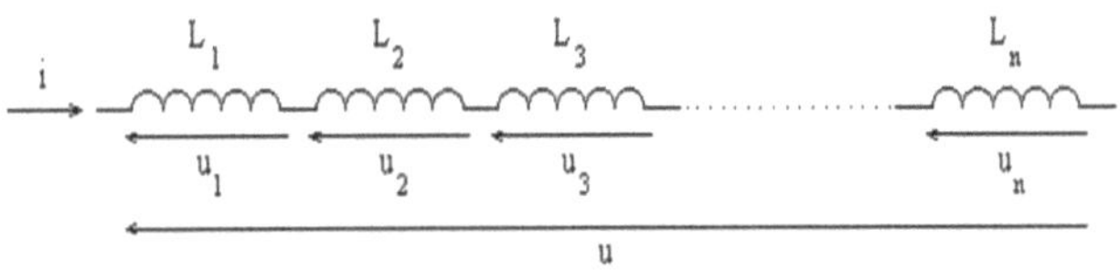

- **Parallel connection of coils:**

$$\frac{1}{L_{eq}} = \sum_{i=1}^{n} \frac{1}{L_i}$$

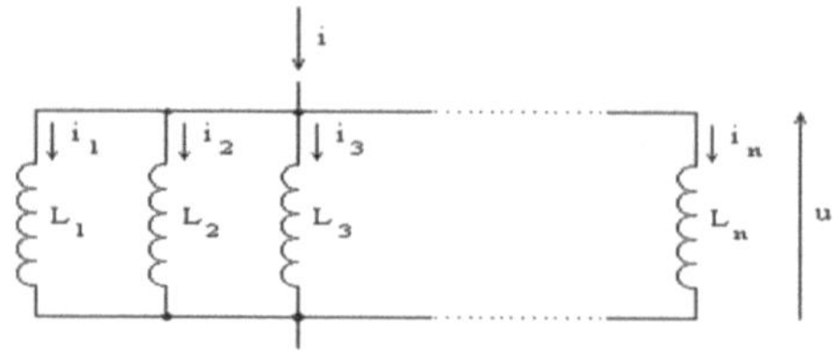

III. Harmonic (sinusoidal) regime

I. Characteristics of a periodic signal

a- Period

The period, denoted by T, of a periodic signal is the smallest duration after which the signal repeats itself identically. Its unit in the International System of Units (SI) is the second: s.

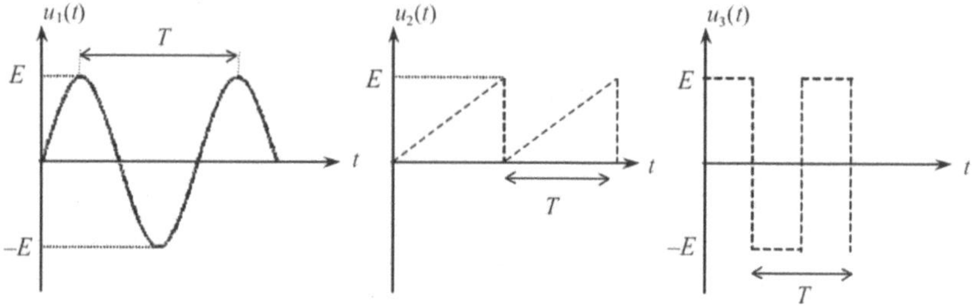

Figure 2.1 Examples of periodic functions with period T

b- Frequency

The frequency, denoted by f, is the number of times an event occurs in one second. Its unit in the SI is the Hertz, denoted as Hz.
Frequency corresponds to the number of periods per second. It is related to the period by the following equation: $f = \frac{1}{T}$

c- Instantaneous voltage

It is given by: $u(t) = U_{\max}\sin(\omega t + \varphi)$,
where :

- u(t) : valeur instantanée de la tension,
- U_{max} : valeur maximale de la tension, en volts.

- ω : The angular frequency of the voltage (also sometimes called angular velocity), in radians per second. The angular frequency is related to the frequency and the period T by: $\omega = 2\pi f = \frac{2\pi}{T}$
- φ : Phase of the voltage at the initial instant, in radians.
- $\omega t + \varphi$: Phase of the voltage at time t, in radians (or in degrees).

d- Valeur moyenne

The average value U_{moy} of any periodic quantity u(t) is calculated using the following relation:

$$U_{moy} = \frac{1}{T}\int_T u(t)dt$$

Note:

The average value of a sinusoidal alternating signal is zero.

e- Effective value or RMS value (Root Mean Square value)

The effective value U_{eff}, of any periodic quantity u(t) is calculated using the following relation:

$$u_{eff} = U = \sqrt{\frac{1}{T}\int_T u^2(t)dt}$$

Note :

The effective value (RMS) of a **sinusoidal alternating signal** is: $u_{eff} = U = \frac{U_{max}}{\sqrt{2}}$

II. Representation of a sinusoidal alternating signal:

a- Cartesian representation

The Cartesian representation uses sinusoidal functions of time.

The following two quantities are defined as follows:

$$u(t) = U_{max}\sin(\omega t + \varphi_u)$$
$$v(t) = U_{max}\sin(\omega t + \varphi_v)$$

They are represented in time alignment as shown below.

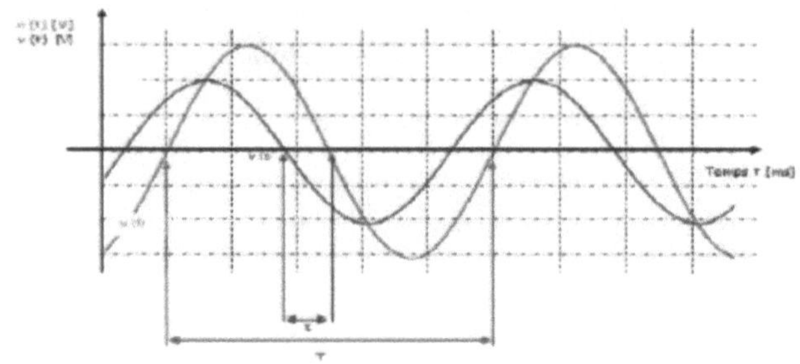

Figure 2.2 Cartesian representation of sinusoidal quantities

The two quantities u(t) and v(t) do not intersect the time axis at the same instant. They are time-shifted. This time shift, denoted as, always has the same value; the quantities u(t) and v(t) do not have the same phase at the initial instant. They are out of phase. τ

b- Vector Representation

When two sinusoidal voltages u(t) and v(t) are being studied, they can be represented using vectors based on their characteristics, as for:

$$u(t) = U_{\max} \sin(\omega t) \ \ (\varphi_u = 0)$$
$$v(t) = U_{\max} \sin(\omega t + \varphi_v)$$

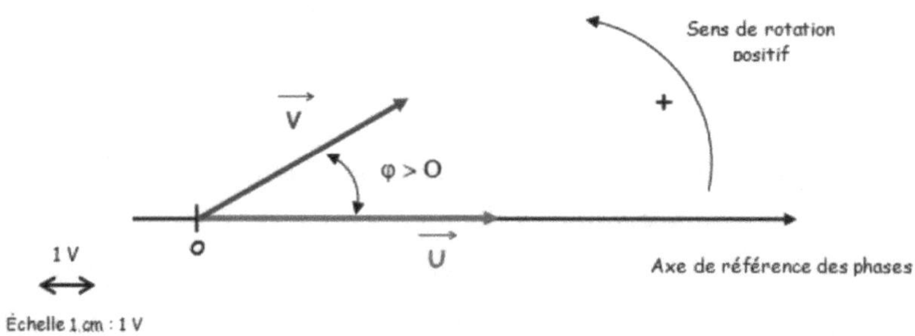

Figure 2.4 Fresnel representation, v(t) leading u(t)

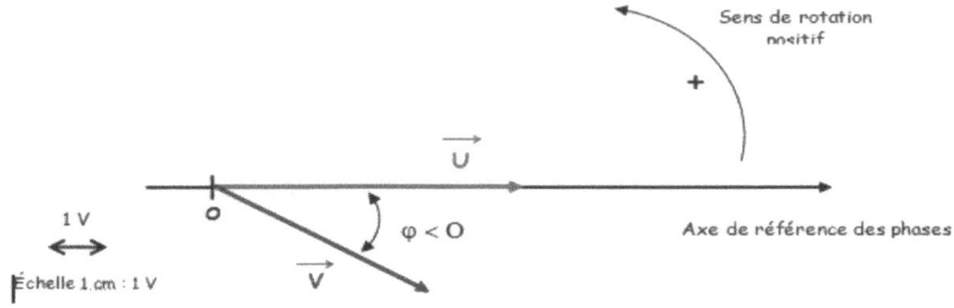

Figure 2.3 Fresnel representation, v(t) lagging u(t)

III. Study of R, L, and C circuits in sinusoidal regime

3-1 Resistif Circuit

Consider a generator producing a sinusoidal voltage with a peak value of 100 V. In a resistive circuit, the current is in phase with the voltage.

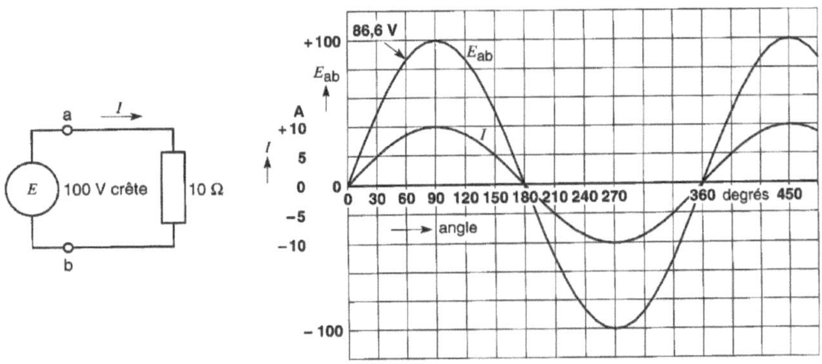

Figure 2.5 Resistive AC Circuit: Current and Voltage Waveform

By studying the waveform of the applied voltage and the resulting current, we can see that both pass through zero at the same time and reach their maximum positive value simultaneously. It is then said that the voltage and the current are in phase.

- **Power dissipated in a resistor**

The power P dissipated by Joule effect in a resistor is given by the product of the voltage E across its terminals and the current I flowing through it. The same formula can be applied to the circuit from the previous figure by multiplying the voltage values by the corresponding current values at each instant, allowing us to plot the variation curve of power P.

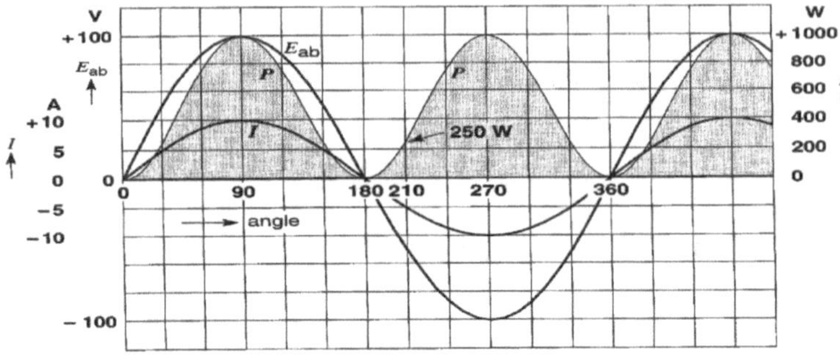

Figure 2.6 Resistive AC Circuit: Current, Voltage and Power Waveform

3.2 Capacitive Circuit

A sinusoidal voltage applied to a capacitive circuit produces a sinusoidal current. However, this capacitive current is 90° ahead of the voltage in phase.

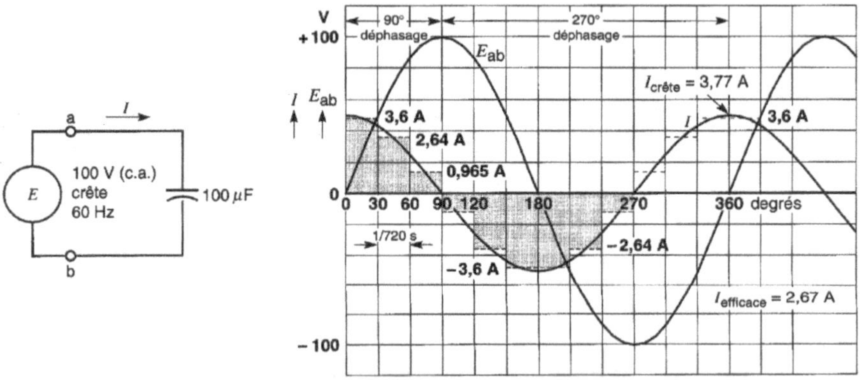

Figure 2.7 Capacitive AC Circuit: Current and Voltage Waveform

- **Capacitive reactance**

The capacitive reactance of a capacitor is given by the formula:

$$X_c = \frac{1}{\omega C} = \frac{1}{2\pi f C} \ [\Omega]$$

f : The frequency of the source in Hertz $[Hz]$

C : The capacitance of the capacitor $[F]$

➤ **Reactive power in a capacitor: The capacitive VAR**

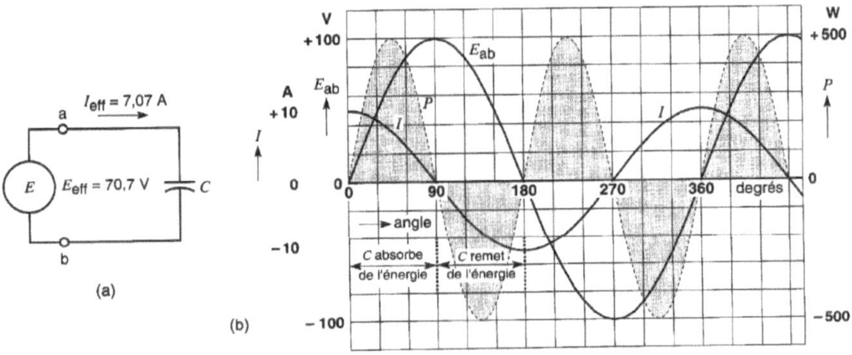

Figure 2.8 Resistive AC Circuit: Current, Voltage and Power Waveform

The product of the effective value of the current in a capacitor and the effective value of the voltage applied across its terminals is called capacitive reactive power: Q_c
The unit of capacitive reactive power is the ***volt-ampere reactive*** (VAR).

This reactive power does not produce any heat dissipation in the capacitor.

3.3 Inductive circuit

A sinusoidal voltage applied to an inductive circuit produces a sinusoidal current. However, this inductive current is 90° behind (lagging) the voltage.

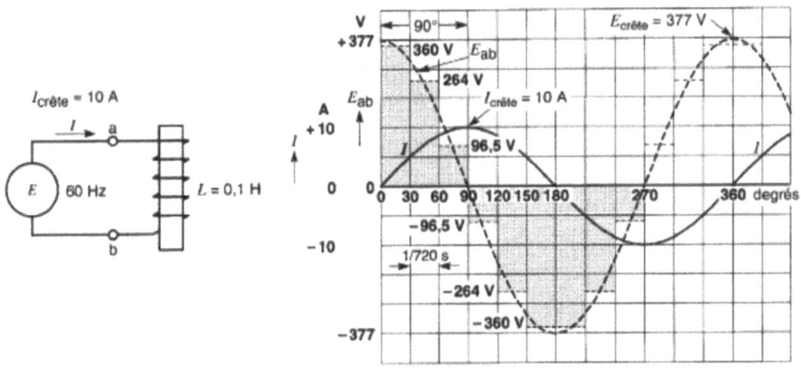

Figure 2.9 Inductive AC Circuit: Current and Voltage Waveform

- **Inductive reactance**

The inductive reactance of a coil is given by the formula: $X_L = L\omega = 2\pi fL \; [\Omega]$

f : The frequency of the source in Hertz $[Hz]$

L: The inductance of the coil $[H]$

- **Recactive power in a coil : Inductive VAR**

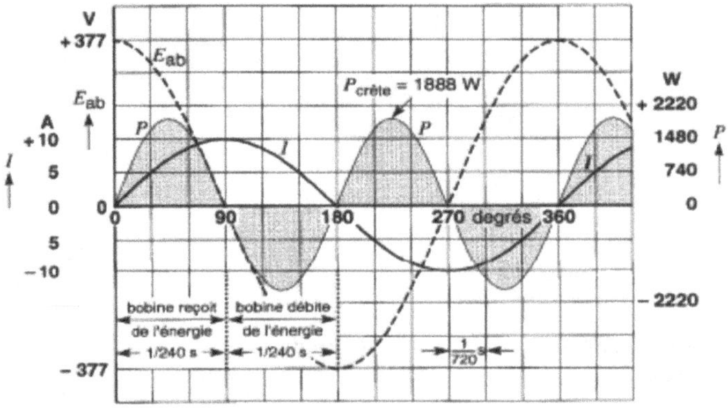

Figure 2.10 Inductive AC Circuit: Current, Voltage and Power Waveform

It is observed that during the first quarter of the period, the power is positive, and the coil receives energy from the source. During the next quarter of the period, the power is negative, and the coil returns the energy it stored back to the source.

Thus, there is a continuous exchange of energy between the source and the coil. The product of the effective value of the current in a coil and the effective value of the voltage applied across its terminals is called ***inductive reactive power*** Q_L.

The unit of reactive power is the **volt-ampere reactive (VAR).**

IV. Transient regime

1- Series R-L Circuit

In the circuit below, the loop rule is written as:

$$-e + Ri + L\frac{di}{dt} = 0. \Rightarrow \quad \frac{di}{dt} + \frac{R}{L}i = \frac{e(t)}{L} \quad \text{(E)}$$

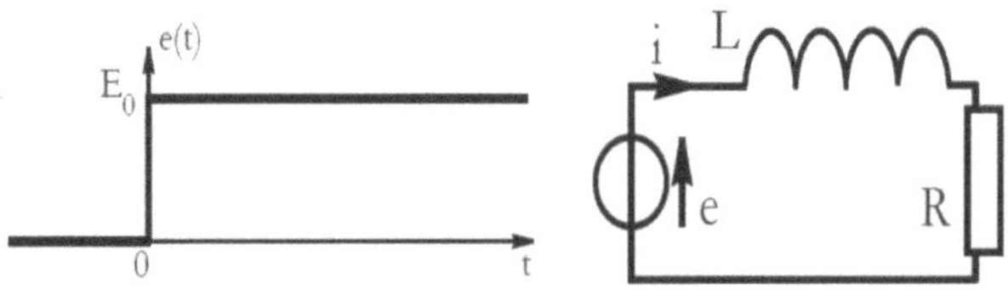

Figure 2.11 Inductive AC Circuit : Transient regime

It is a first-order linear differential equation with constant coefficients and a second member.

e(t) is a voltage step, such that: $\begin{cases} t<0: \text{ e(t)=0} \\ t\geq 0: \text{ e(t)=E}_0 \end{cases}$ $\text{A}\, t \geq 0,$

The differential equation is written as: $\dfrac{di}{dt}+\dfrac{R}{L}i=\dfrac{E_0}{L}$ (1)

After solving equation (1), assuming all initial conditions are zero, we obtain:

$$i(t)=\frac{E_0}{R}(1-e^{-\frac{R}{L}t})$$

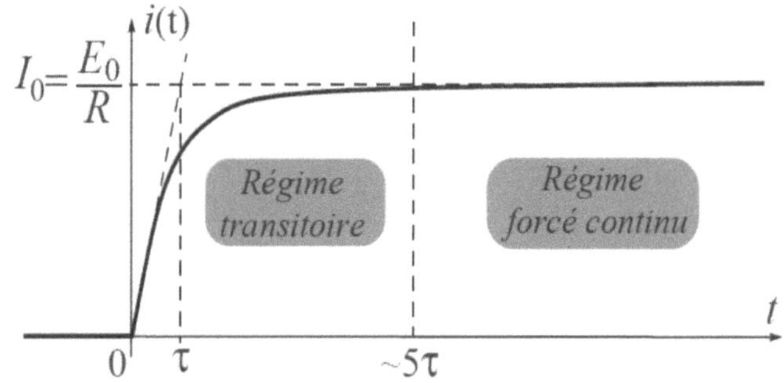

Figure 2.12 :Transient and Steady-State Response of an RL Circuit

$\tau=\dfrac{L}{R}$ is called the time constant of the series RL circuit. It provides an estimate of the duration of the transient regime.

Whene $t\to\infty$, $\text{i(t)}\to\dfrac{E_0}{R}=I_0$: **The transient regime** fades away and is replaced by the **steady-state DC regime.**

Example : $\begin{cases} L = 1mH \\ R = 1K\Omega \end{cases} \Rightarrow \tau = \frac{L}{R} = \frac{10^{-3}}{10^{3}} = 10^{-6}s = 1\mu s$

It is quite small: the transient regime fades away fast.

- **Representation of the voltage across the coil's terminals U_L:**

$$U_L = L\frac{di}{dt} = L\frac{E_0}{L}e^{-\frac{R}{L}t} = E_0 e^{-\frac{R}{L}t} = E_0 e^{-\frac{t}{\tau}}$$

During the transient regime, the coil attempts to oppose the generator's voltage by generating a voltage in the opposite direction, in accordance with **Lenz's law**.

In the steady-state (DC steady-state), U_L equals zero.

In the DC regime, the coil functions as a conductive wire.

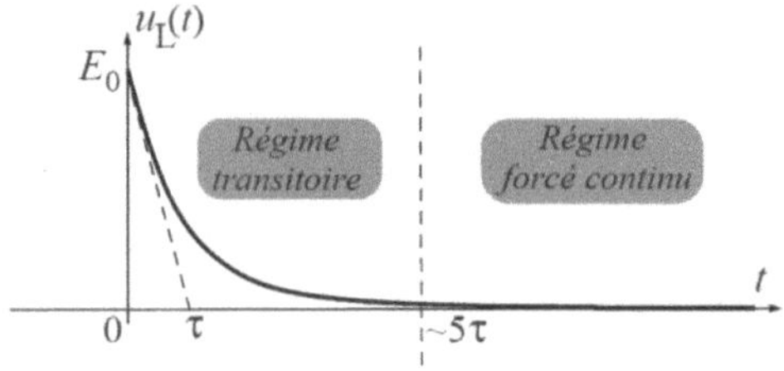

Figure 2.13: Transient and Steady-State representation of U_L

- **Extinction of the source (study of the natural response)**

To simplify the calculations, we reset the time:

$\begin{cases} t < 0: \; e(t) = E_0 \\ t \geq 0: \; e(t) = 0 \end{cases}$

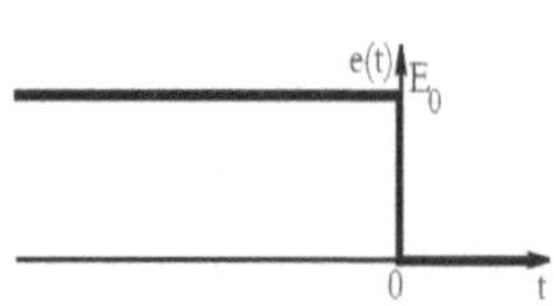

The circuit then reduces to:

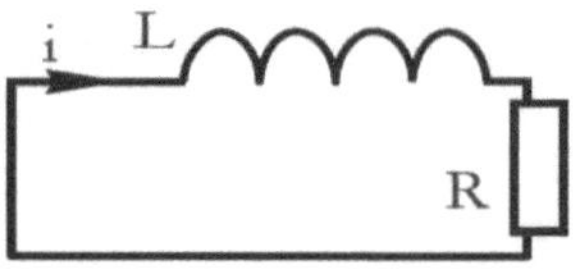

The loop rule is written for $t \geq 0$: $L\frac{di}{dt} + Ri = 0$

It is a first-order linear differential equation with constant coefficients and no second member.

The solution is written as follows: $Be^{-\frac{R}{L}t}$ with $B \in R$

Moreover, due to the continuity of the current flowing through the coil, we have:

$$i(0^+) = \begin{cases} i(0^-) = I_0 = \dfrac{E_0}{R} \\ i(t = 0^+) = B \end{cases} \Rightarrow B = \frac{E_0}{R}, \text{ Then: } i(t) = \frac{E_0}{R} e^{-\frac{R}{L}t}$$

Thus, the voltage across the coil is:

$$u_L(t) = L\frac{di}{dt} = L\frac{E_0}{R}(-\frac{R}{L}e^{-\frac{R}{L}t}) \Rightarrow u_L(t) = -E_0 e^{-\frac{R}{L}t}$$

It is evident that the free response is a transient regime with a limited duration.

After a few values of τ, $i \to 0$ et $u_L \to 0$

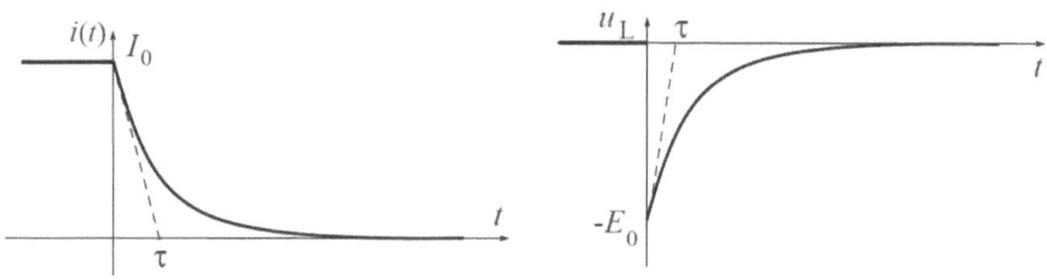

Figure 2.14 : I_L and U_L curves after extinction of the source

Experimental study

A Low Frequency Generator (LFG) delivers a square wave signal with a period T.

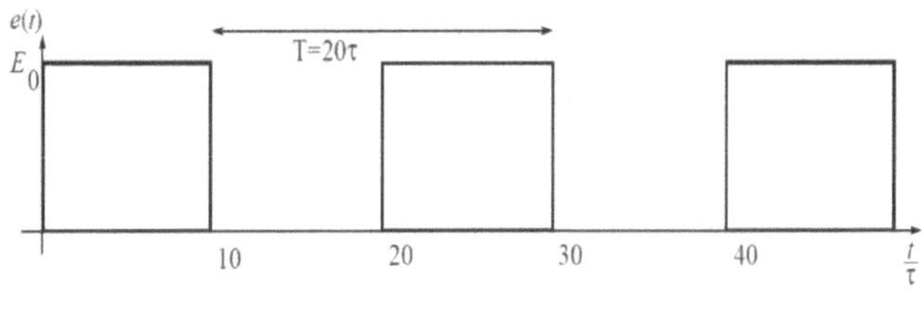

Figure 2.15: Square wave signal

Obtained curves:

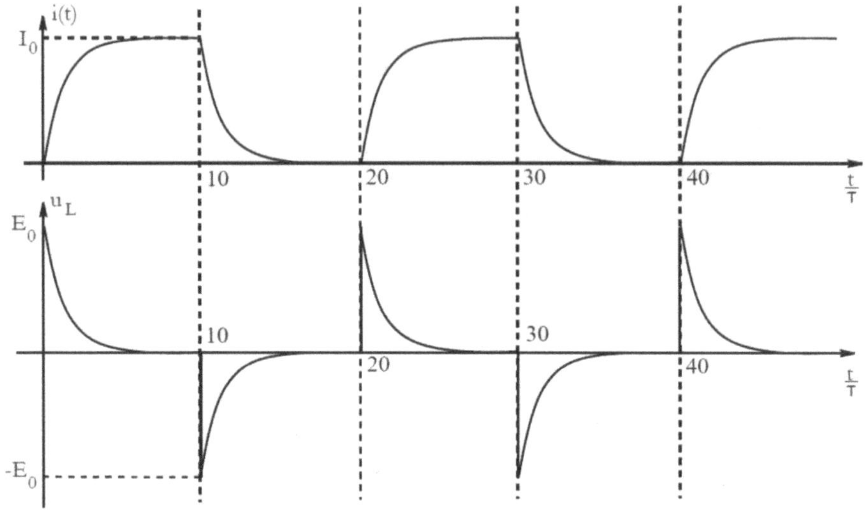

Figure 2.16: Obtained curves of I_L and U_L

2- Series R-C Circuit

➢ **Charging of the capacitor**

In the circuit depicted, the loop rule is expressed as

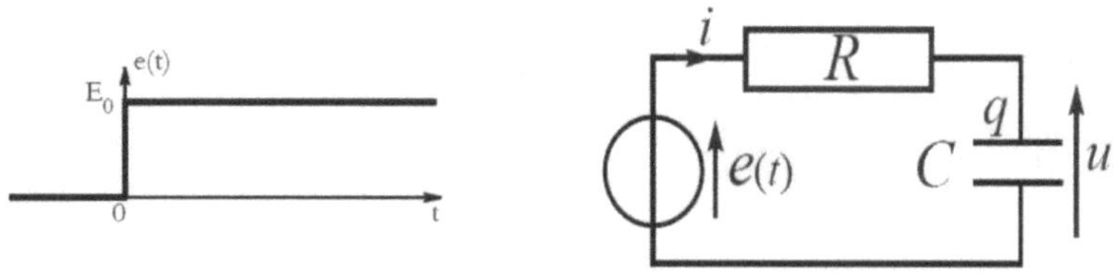

Figure 2.17 Capacitive AC Circuit: Transient regime

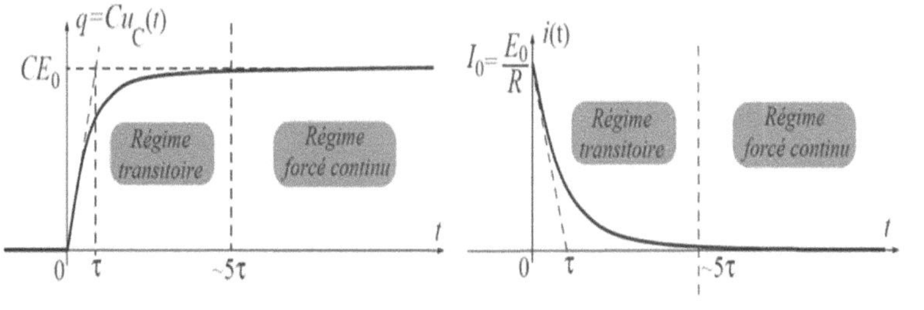

Figure 2.18: Transient and Steady-State response of an RC Circuit

$-e+Ri+u=0.$ (1) with $u=\frac{1}{C}\int idt \Rightarrow i=C\frac{du}{dt}$

$$(1)\Rightarrow -e+RC\frac{du}{dt}+u=0\Rightarrow \frac{du}{dt}+\frac{1}{RC}u=\frac{e}{RC}$$

With τ=RC is the time constant of the series RC circuit

It is a first-order linear differential equation with constant coefficients with a second member.

e(t) is a step voltage, such that: $\begin{cases} t<0: \text{ e(t)=0} \\ t\geq 0: \text{ e(t)=E}_0 \end{cases}$

For $t\geq 0$, The differential equation is written as: $\frac{du}{dt}+\frac{1}{\tau}u=\frac{E_0}{\tau}$ (2)

The constant particular solution corresponds to the steady-state regime, so: $\mathbf{u_0(t) = E_0}$
The corresponding homogeneous equation has the general solution: $\mathbf{u_1(t) = A\, e^{-t/\tau}}$

The solution of equation (2) is therefore: $u(t)=E_0+Ae^{-\frac{t}{\tau}}$

To determine A, we assume that for t < 0, the capacitor is uncharged

$(u(t = 0^-) = u_0 = 0)$.

Furthermore, the continuity of the charge on the capacitor's plates requires that:

$u(t = 0^+)= u(t = 0^-)$.

Then : $u(t=0)=E_0+A=0\Rightarrow A=-E_0$

Finally, the solution of equation 2 is: $u(t) = E_0(1 - e^{-\frac{t}{\tau}})$ avec τ=RC

We can then deduce the current through the capacitor C, which is given by:

$$i(t) = C\frac{du(t)}{dt} = C\frac{d(E_0(1 - e^{-\frac{t}{RC}}))}{dt} \quad \Rightarrow \quad i(t) = \frac{E_0}{R}e^{-\frac{t}{RC}}$$

By setting $I_0 = \frac{E_0}{R}$, we will have: $\boxed{i(t) = I_0 e^{-\frac{t}{RC}}}$

Notes:

1- The steady-state is reached when the capacitor has reached its maximum charge under the voltage E0; at that point, no current flows.

2- In steady state, the capacitor behaves like an **open switch.**

➢ **Discharge of the capacitor**

To simplify the calculations, we reset the time:

$$\begin{cases} t < 0: \ e(t) = E_0 \\ t \geq 0: \ e(t) = 0 \end{cases}$$

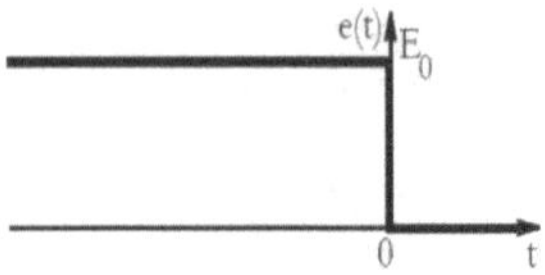

The circuit is then reduced to:

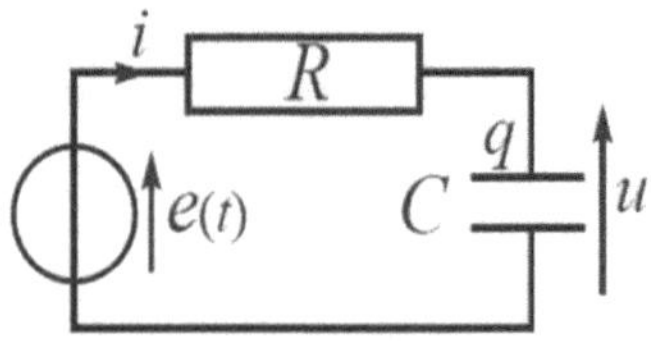

The loop rule is then written as:

$$Ri + u = 0.(3) \quad with \quad u = \frac{1}{C}\int idt \quad \Rightarrow i = C\frac{du}{dt}$$

$$(3) \Rightarrow RC\frac{du}{dt} + u = 0 \Rightarrow \frac{du}{dt} + \frac{1}{RC}u = 0$$

With τ=RC is the time constant of the series RC circuit

It is a first-order linear differential equation with constant coefficients with no second member.

The general solution of equation (3) is of the form : $u(t) = Be^{-\frac{t}{RC}}$

Since the voltage u(t) is continuous, we have u(t=0) = E0, consequently:

$$u(t) = E_0 e^{-\frac{t}{RC}}$$

We can deduce the capacitor's discharge current as:

$$i(t) = C\frac{du(t)}{dt} = CE_0\frac{d}{dt}(e^{-\frac{t}{RC}}) = -\frac{E_0}{R}e^{-\frac{t}{RC}}$$

By setting $I_0 = -\frac{E_0}{R}$, we will have: $i(t) = -I_0 e^{-\frac{t}{RC}}$

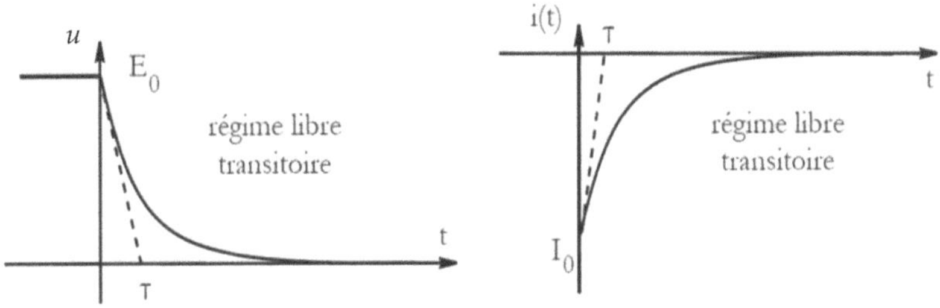

Figure 2.19 : I_C and U_C curves after extinction of the source

Remarque

- $i < 0$ because the discharge occurs in the opposite direction to the conventional positive direction of the diagram.

Experimentale study

A Low Frequency Generator (LFG) delivers a square wave signal with a period T.

Obtained Curves

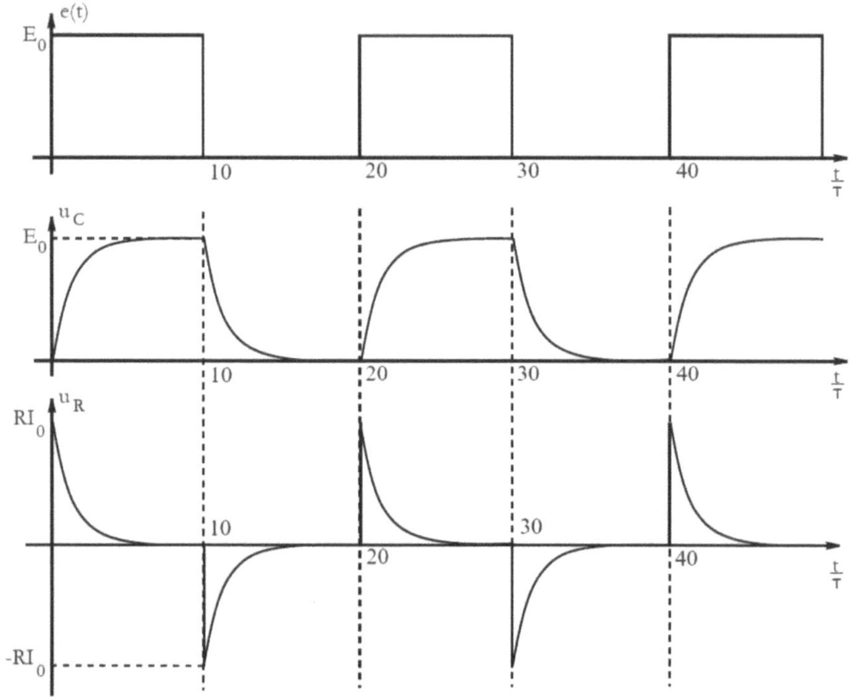

Figure 2.20: Obtained curves of U_C and U_R

Notes :

It is observed that the voltage u is a continuous function of time, while the current i experiences discontinuities.

V. Conclusion

This chapter provided a comprehensive review of the fundamental laws governing electrical circuits. It covered key concepts in direct current (DC), harmonic (sinusoidal) regimes, and transient analysis, laying the groundwork for understanding more advanced electrical systems.

In the DC regime, circuit analysis is simplified using Ohm's law and Kirchhoff's laws, with resistors, capacitors, and inductors exhibiting distinct steady-state behaviors.

The sinusoidal regime introduced the importance of frequency-dependent elements such as reactance and impedance, demonstrating how alternating currents interact with resistive, capacitive, and inductive components.

The transient regime explored the time-dependent response of RL and RC circuits, highlighting the importance of time constants in defining how quickly a system reaches steady-state conditions.

These fundamental laws form the basis for analyzing and designing electrical circuits, particularly in applications involving signal processing, power systems, and control engineering. A strong grasp of these principles is essential for tackling more complex electrical and electronic systems.

Electric circuits and power

Table of contents

I- Introduction

In physics, power represents an amount of energy per unit of time. Its unit is the Watt (1 W = 1 J/s). Generally, the power driving energy conversion systems is the average power of these systems, also referred to as *active power*. In electrical systems, power is a critical parameter that determines the performance and efficiency of energy conversion systems, such as motors, generators, and electronic devices.

An electrical dipole under an effective voltage V and carrying an effective current I consumes a power $P = k.V.I$

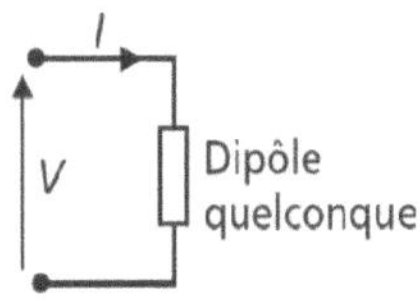

k: power factor, ranging between 0 and 1.

II- Electric Power in DC Systems

Electric power in direct current (DC) systems is a fundamental concept that applies to many applications, from simple circuits to advanced power systems. In a DC circuit, power is calculated by the product of the voltage across a component and the current flowing through it.

The simplicity of DC power calculations makes it ideal for use in low-voltage electronic devices, battery-powered systems, and renewable energy sources such as solar panels. Additionally, since DC power remains constant over time, there are no phase shifts or reactive components to consider, unlike AC power systems.

This allows for efficient transmission and utilization in various applications, including electric vehicles and portable electronic devices.

Understanding DC power principles is essential for designing reliable electrical circuits and optimizing energy consumption in modern technology.

The DC regime represents the simplest case for calculating electrical power since *k*=1. The only passive load being the resistance, this calculation can be summarized in the diagram below:

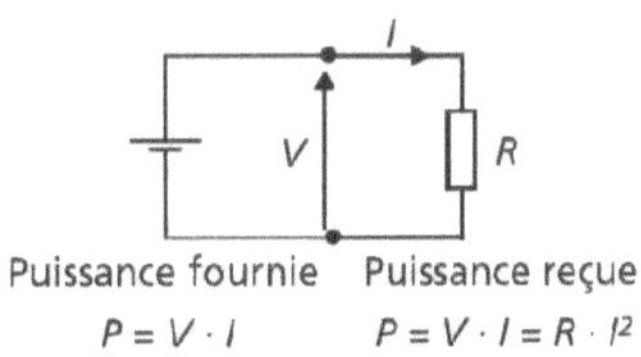

III- Electrical power in sinusoidal alternating current (AC) regime

In the sinusoidal alternating current (AC) regime, the focus is on the average power consumed by electrical loads over a complete cycle.

Unlike in DC circuits, where power remains constant, AC power fluctuates due to phase differences between voltage and current.

To better understand the power dynamics, we analyze different components of power. Considering an inductive dipole subjected to a sinusoidal voltage and current, we distinguish several types of power, each serving a specific role in energy transfer and system behavior.

These power components define the efficiency and performance of AC electrical systems, impacting both energy consumption and operational stability.

We focus on the general case of an inductive dipole subjected to a sinusoidal voltage $v(t) = V\sqrt{2}\sin(\omega t)$ and traversed by a current $i(t) = I\sqrt{2}\sin(\omega t - \varphi)$.

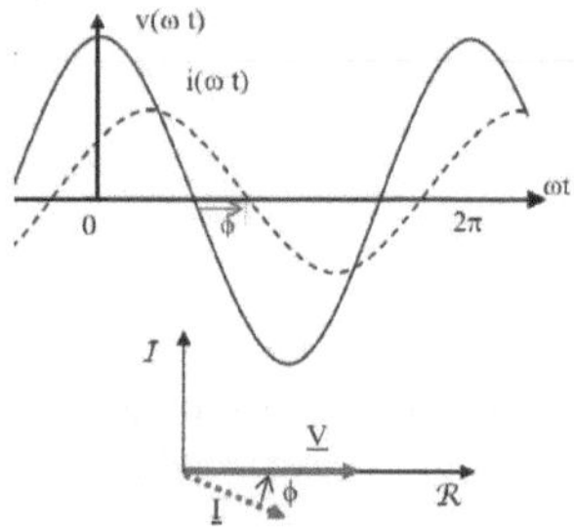

We then distinguish the following types of power:

- **Instantaneous power:**

It is the product of current and voltage at each instant:

$$p(t) = v(t).\, i(t) = V\sqrt{2}\sin(\omega t).\, I\sqrt{2}\sin(\omega t - \varphi)$$

Reminder: $\cos(a+b) = \cos a\cos b - \sin a \sin b$ et $\cos(a-b) = \cos a \cos b + \sin a \sin b$

After simplifying the product, we obtain:

$$p(t) = VIcos(\varphi) - VIcos(2\omega t - \varphi)$$

- **Fluctuanting Power :**

This is the variable part of the instantaneous power: $VIcos(2\omega t - \varphi)$

- **Active power :**

This is the average value of the instantaneous power:

$$P = \prec p(t) \succ = VIcos(\varphi)$$

This is the power that corresponds to actual physical work, and its unit is the *Watt*(W).

- **Apparent power.**

It is the product of the RMS (Root Mean Square) values (effective):

$$S = V.I$$

This power is often called "rated power."
It is the characteristic quantity for the insulation and the cross-sectional area of the conductors, i.e., the dimensions of the equipment.
Its unit is the *Volt-Ampere (VA).*

Note:

In sinusoidal alternating current (AC) regime (only), the power factor is defined as the quantity (dimensionless): $k = F_p = \frac{P}{S} = \cos(\varphi)$

- **Reactive power.**

It is the power with no physical effect in terms of work, corresponding to the "reactive" part of the current. It is defined only in the sinusoidal regime and is expressed as:

$$Q = VIsin(\varphi)$$

Its unit is the Volt-Ampere Reactive (VAR).

Relationships between P, Q est S

The relationships between active power (P), reactive power (Q), and apparent power (S) in a sinusoidal AC system are given by:

$$S = \sqrt{P^2 + Q^2}$$

Knowing that: $P = VIcos(\varphi)\ et\ Q = VIsin(\varphi)$

This formulation reveals a graphical relationship between the different quantities. This is referred to as the power triangle:

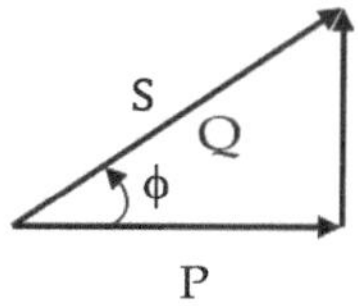

On the other hand, the characteristic quantity is also revealed : $tg(\varphi)=\dfrac{Q}{P}$

Summary:

Once these powers (active, reactive, and apparent) are defined, it is essential to memorize their definitions and the relationships between them.
Active power (P) represents the useful power consumed by the load, reactive power (Q) accounts for the energy stored and returned by inductive or capacitive components, and apparent power (S) is the total power supplied by the source:

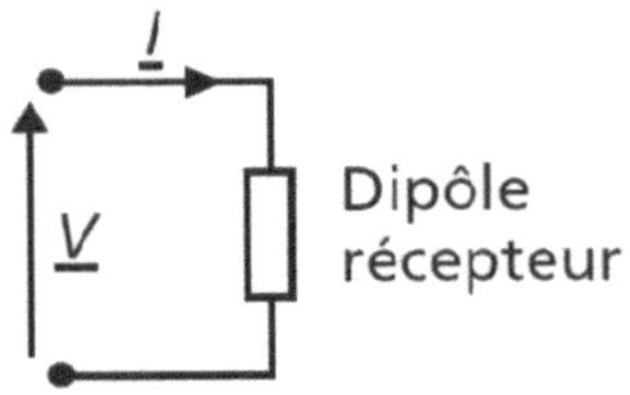

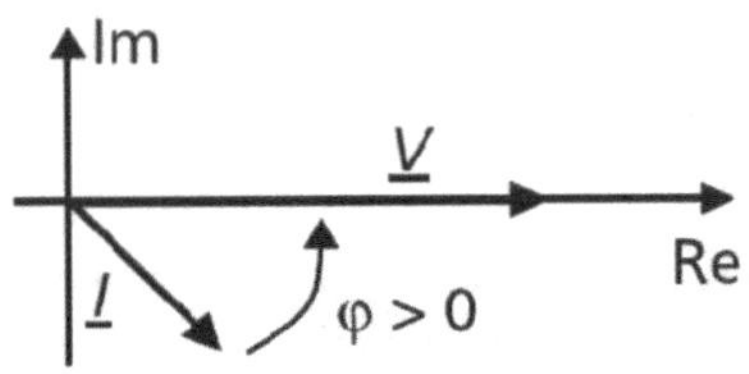

- Active power : $P=VIcos(\varphi)(W)$
- Reactive power : $Q=VI\sin(\varphi)(VAR))$
- Apparent power : $S(VA)=VI=\sqrt{P^2+Q^2}$
- In sinusoidal regime: $V=V_{eff}=\dfrac{V_{\max}}{\sqrt{2}}$ and $I=I_{eff}=\dfrac{I_{\max}}{\sqrt{2}}$
- Power factor : $k=F_p=\dfrac{P}{S}=\cos(\varphi)$ and $tg(\varphi)=\dfrac{Q}{P}$

➢ Complex apparent power

To analytically determine the various powers, the complex apparent power is formed:

$$\underline{S} = \underline{V}.\underline{I}^*$$

(Where $\underline{I}^*$ is the complex conjugate of $\underline{I}$).

Knowing that: $\underline{I} = Ie^{-j\varphi} = I\cos\varphi - jI\sin\varphi \Rightarrow \underline{I}^* = Ie^{+j\varphi} = I\cos\varphi + jI\sin\varphi$

Then, $\underline{S} = \underline{V}.\underline{I}^* = V.Ie^{+j\varphi} = VI\cos\varphi + jVI\sin\varphi = P + jQ$

Therefore, we have: $\underline{S} = \underline{V}.\underline{I}^* = P + jQ \Rightarrow |\underline{S}| = S = \sqrt{P^2 + Q^2}$

The following table summarizes the active and reactive powers of the most commonly encountered components in electrotechnics, in a sinusoidal alternating current (AC) regime:

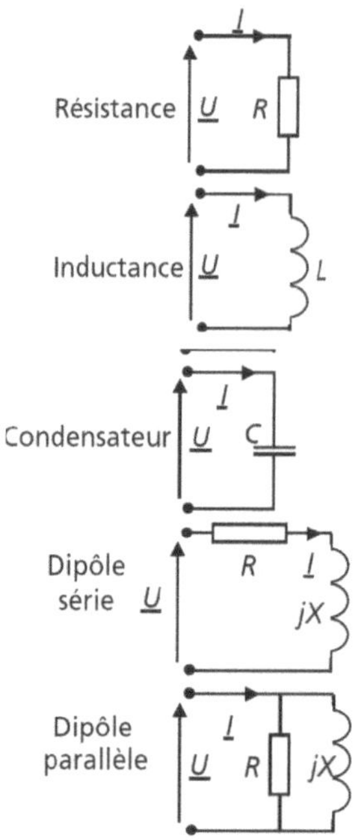

$\underline{S}$	P	Q
$\underline{S} = R \cdot \underline{I} \cdot I^*$ $= RI^2$ $= U^2 / R$	$RI^2 = U^2 / R$	0
$\underline{S} = jL\omega \cdot I \cdot I^*$ $= jL\omega I^2$ $= jU^2 / L\omega$	0	$jL\omega I^2 = jU^2 / L\omega$
$\underline{S} = -j / C\omega \cdot I \cdot I^*$ $= -j / C\omega I^2$ $= jU^2 / C\omega$	0	$-j / C\omega I^2 = -jC\omega U^2$
$\underline{S} = (R + jX)I \cdot I^*$ $= RI^2 + jX \cdot I^2$	$R \cdot I^2$	$X \cdot I^2$
$\underline{S} = \underline{U} \cdot I^*$ $\underline{U} = \underline{I} / (R // jX)$	U^2 / R	U^2 / X

IV- Theorem of Boucherot and the Power Triangle

Theorem of Boucherot.

Boucherot's theorem states that the total reactive power can be found by an arithmetic sum of its components and the total real power can likewise be found by an arithmetic sum of its components. However, this is not true for apparent power.

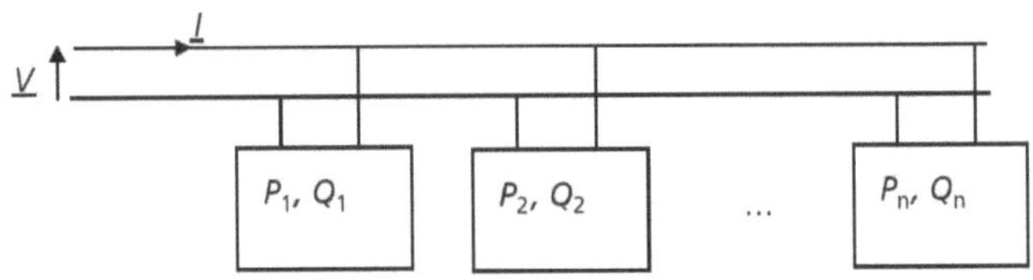

$S = V \cdot I$ $\quad P = P_1 + P_2 + \ldots + P_n$ $\quad Q = Q_1 + Q_2 + \ldots + Q_n$

$\underline{S} = \underline{S}_1 + \underline{S}_2 + \ldots + \underline{S}_n$ mais attention : $S \neq S_1 + S_2 + \ldots + S_n$

Représentation de la conservation des puissances sous la forme de triangles des puissances :

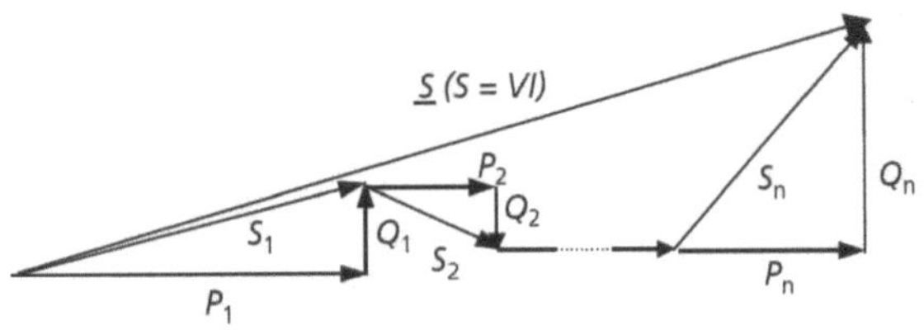

V- Conclusion

This chapter provided an in-depth exploration of electrical power in both DC and AC systems, emphasizing the different types of power and their significance in circuit analysis.

In DC systems, power calculation is straightforward as it only involves voltage and current, making it ideal for applications such as electronic devices and renewable energy systems. In AC systems, power is more complex due to phase differences between voltage and current, leading to the distinction between active power (W), reactive power (VAR), and apparent power (VA).

The power triangle and Boucherot's theorem were introduced as essential tools for understanding power relationships in AC circuits. The power factor was highlighted as a key parameter affecting efficiency and system performance.

Mastering these concepts is essential for designing and optimizing electrical and electronic systems, ensuring efficient power transmission and utilization in various applications.

Three-phase Electrical systems

Table of contents

I. Introduction

The transmission of electrical energy is carried out using three-phase systems because a three-phase line dissipates less electrical energy than a single-phase line for the same amount of power delivered.

Three-phase systems provide a more balanced and efficient distribution of power, reducing losses and improving the overall performance of the electrical grid. Additionally, three-phase systems require less conductor material compared to single-phase systems, making them more cost-effective for long-distance power transmission. This efficiency and reliability make three-phase systems the standard for industrial and large-scale power distribution networks.

The production of electrical energy is carried out using three-phase systems because, for equal power ratings, a three-phase machine is less expensive than a single-phase machine.

This cost efficiency arises from the fact that the cost of electrical machines is directly related to their mass, and at equal power, a single-phase machine is approximately 1.5 times heavier than a three-phase machine.

Three-phase systems also provide a more balanced and efficient power generation process, reducing mechanical vibrations and improving the overall performance of generators. Additionally, three-phase machines are more compact and require less material, making them easier to manufacture and maintain. These advantages make three-phase systems the preferred choice for power generation in industrial and utility-scale applications.

The distribution is carried out using four terminals:

- Three phase terminals labeled as 1, 2, 3 or A, B, C or R, S, T;
- A neutral terminal labeled as N.

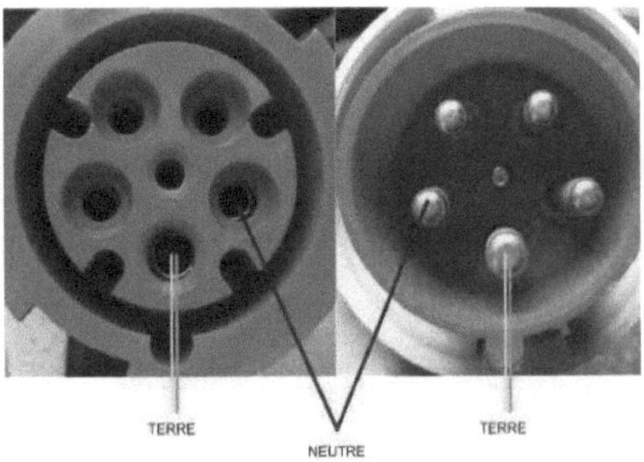

II. Study of Phase Voltages

Balanced three-phase voltages [or currents] are defined as three sinusoidal alternating voltages [or currents] with the same frequency, the same RMS value, and a phase shift of 120° between each pair.

This symmetry ensures that the system remains balanced, with the sum of the three voltages [or currents] at any instant being zero. The 120° phase separation allows for a constant power transfer, minimizing fluctuations and improving the efficiency of the system. Balanced three-phase systems are widely used in power generation, transmission, and distribution due to their stability and ability to deliver consistent power to loads.

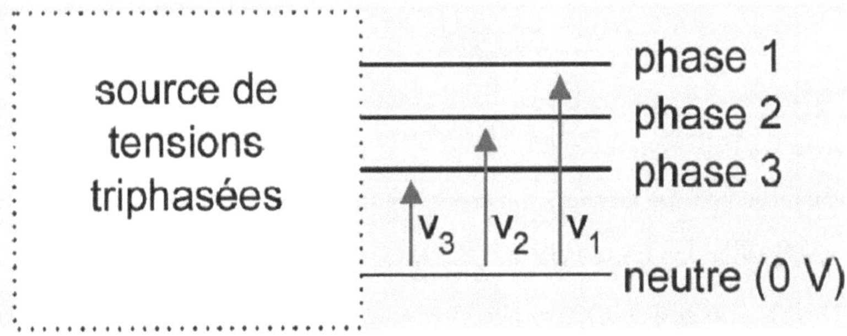

The voltages v_1, v_2, and v_3 are referred to as phase voltages (voltages between a phase and the neutral).

Time Domain Representation

Choose v_1 (t) as the reference voltage :

$$v_1(t) = V\sqrt{2}\,sin(\omega t)$$

$$v_2(t) = V\sqrt{2}\,sin(\omega t - \frac{2\pi}{3}) \text{ avec } -\frac{2\pi}{3} = -120^\circ$$

$$v_3(t) = V\sqrt{2}\,sin(\omega t - \frac{4\pi}{3}) \text{ avec } -\frac{4\pi}{3} = -240^\circ$$

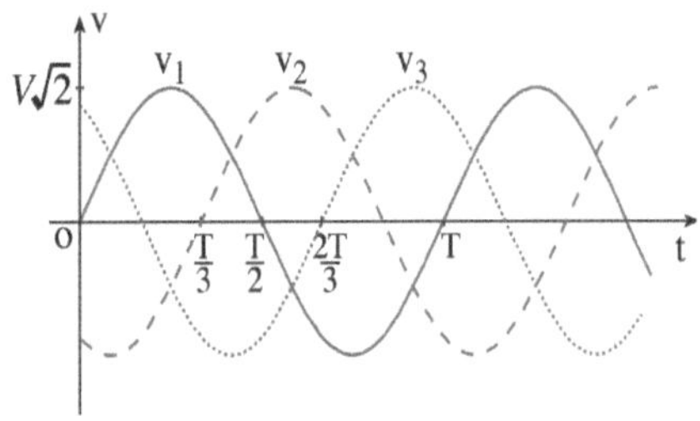

Fresnel representation

From the previously mentioned time-dependent equations, the following vectors are derived:

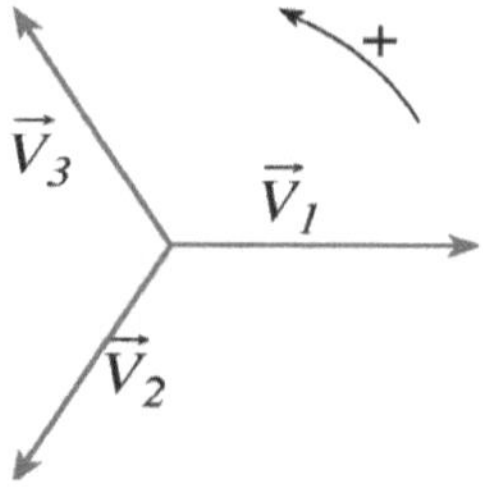

$$\vec{V}_1\begin{pmatrix} V \\ 0 \end{pmatrix} = V\angle 0;$$

$$\vec{V}_2\begin{pmatrix} V \\ -\frac{2\pi}{3} \end{pmatrix} = V\angle -\frac{2\pi}{3};$$

$$\vec{V}_3\begin{pmatrix} V \\ -\frac{4\pi}{3} \end{pmatrix} = V\angle -\frac{4\pi}{3}$$

The system is a balanced direct system.

- Balanced because Fresnel's construction shows that:

  $\vec{V}_1 + \vec{V}_2 + \vec{V}_2 = \vec{0} \Rightarrow v_1 + v_2 + v_3 = 0$

- Direct because a stationary observer would see the vectors passing in front of them in the order 1, 2, 3r.

III. Study of composed voltages.

The voltages u_{ij} are called composed voltages (or phase-to-phase voltages). The effective value of the phase-to-phase voltages is denoted by U.

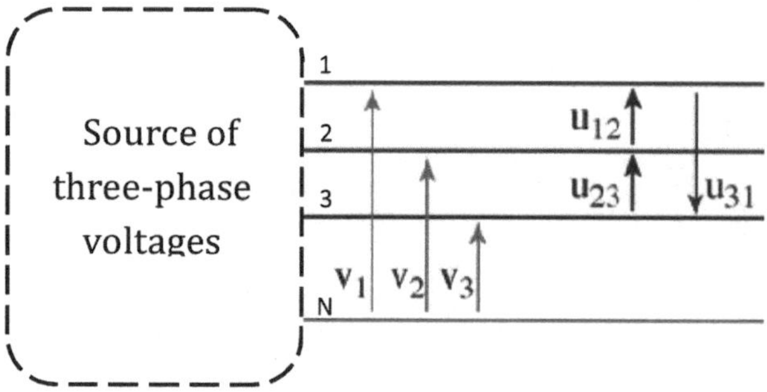

The composed voltages have the same frequency as the simple voltages.

$$u_{12} = v_1 - v_2 \Rightarrow \overline{U}_{12} = \vec{V}_1 - \vec{V}_2$$

$$u_{23} = v_2 - v_3 \Rightarrow \overline{U}_{23} = \vec{V}_2 - \vec{V}_3$$

$$u_{31} = v_3 - v_1 \Rightarrow \overline{U}_{31} = \vec{V}_3 - \vec{V}_1$$

- **Associated Fresnel Vectors**

$$\vec{U}_1 = \left(U, \frac{\pi}{6}\right),$$

$$\vec{U}_2 = \left(U, -\frac{3\pi}{6}\right),$$

$$\vec{U}_3 = \left(U, -\frac{7\pi}{6}\right)$$

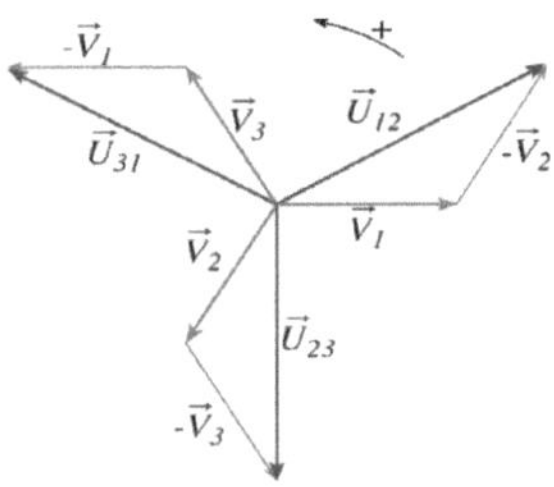

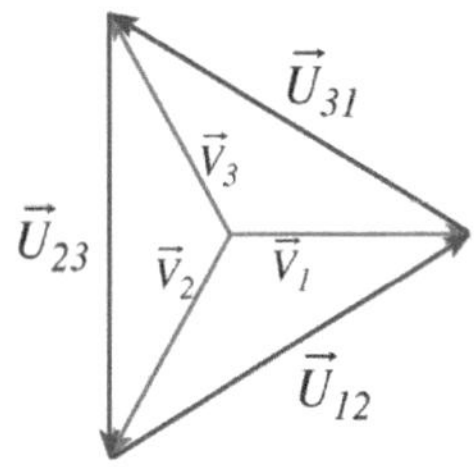

If the network is balanced:

$$\vec{U}_{12} + \vec{U}_{23} + \vec{U}_{31} = \vec{0} \Rightarrow u_{12} + u_{23} + u_{31} = 0$$

The system of three-phase voltages is balanced and in the direct sequence.

- **Time-domain equations and oscillograms**

$$u_{12}(t) = U\sqrt{2}\sin(\omega t + \frac{\pi}{6})$$

$$u_{23}(t) = U\sqrt{2}\sin(\omega t - \frac{\pi}{2})$$

$$u_{31}(t) = U\sqrt{2}\sin(\omega t - \frac{7\pi}{6})$$

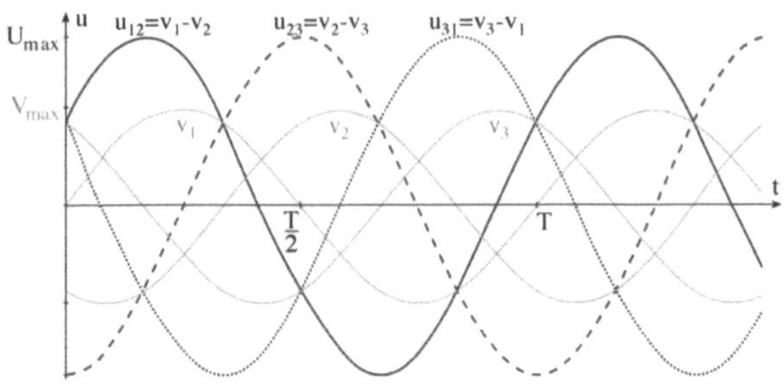

- **Relationship between U and V :**

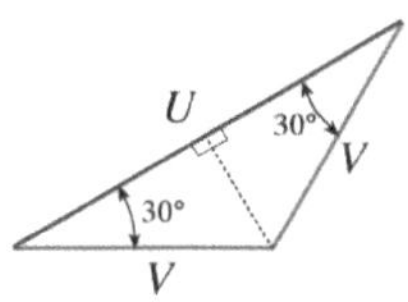

$$U = 2V\cos 30^o \text{ soit } U = 2V\frac{\sqrt{3}}{2}$$

Finaly : $\boxed{U = V\sqrt{3}}$

This relationship is always valid regardless of the load.

Note:

- ✓ In Algeria, SONELGAZ provides a three-phase network.

- 380 V (RMS value between phases)
- 50 Hz
- RMS value of phase voltages:

$$V = \frac{U}{\sqrt{3}} = \frac{380}{\sqrt{3}} \approx 220V$$

In your case, the single-phase voltage (the "mains") comes from a three-phase network where the neutral is used with one of the three phases.

IV. Balanced three-phase loads

In single-phase, the load is a dipole. One terminal is connected to the neutral, and the other to the phase:

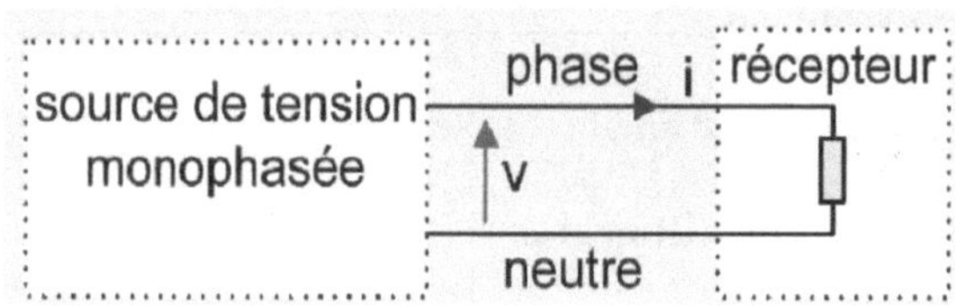

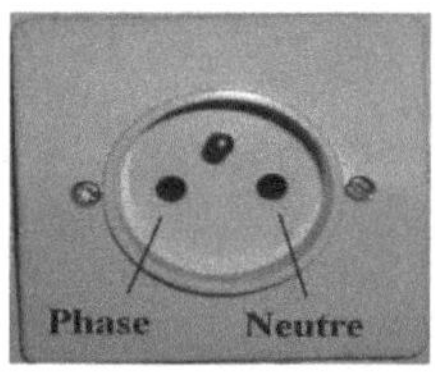

In three-phase, the load has three terminals (one for each phase) and possibly a fourth for the neutral:

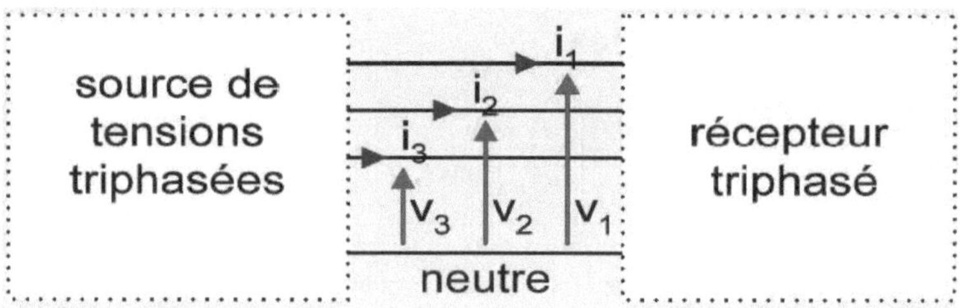

he currents i_1, i_2, and i_3 are called line currents.

IV-1- Star (Y) connection of a three-phase load

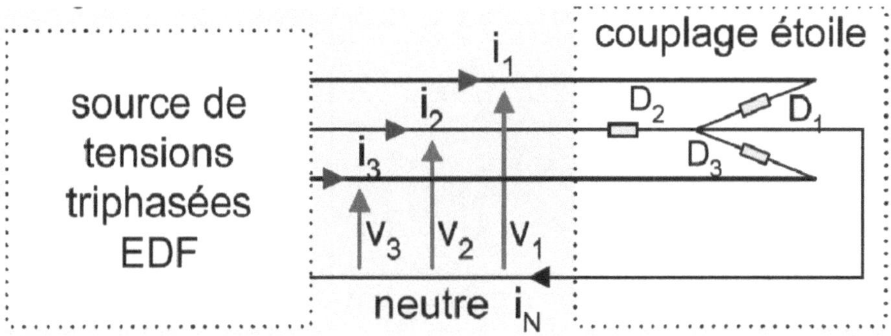

A three-phase load is balanced if it consists of three identical dipoles. Otherwise, it is referred to as an unbalanced three-phase load.
Therefore, in a linear and balanced load, the line currents form a three-phase current system (with the same RMS values III and phase shifts of 120°).

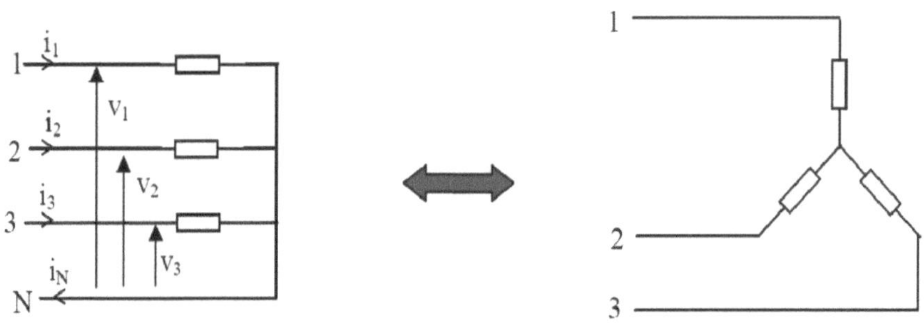

Each phase of the load is subjected to a phase voltage and is traversed by a line current $i=j$.

Therefore, we have: $I_1 = I_2 = I_3 = I$ and $I = J = V / Z$

The load is a balanced three-phase load, and the system of phase voltages is also balanced.

Therefore: $I_1 = I_2 = I_3 = 0$ (Because: $V1 / Z + V2 / Z + V3 / Z = 0$); So $I_N = 0$

Conclusion :

When the installation is balanced, the current in the neutral wire is zero.

- **Fresnel Representation**

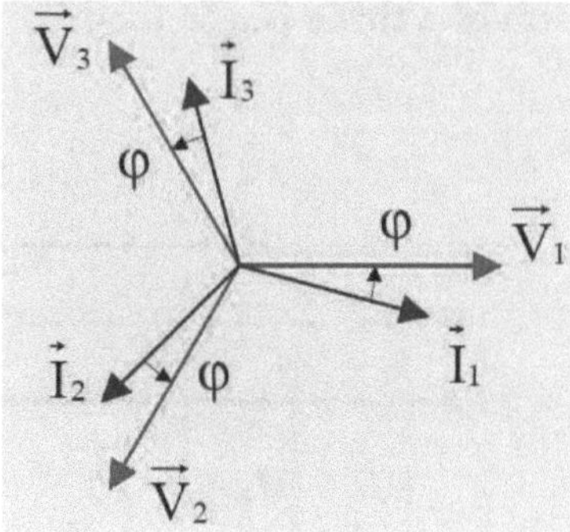

Note: In practice: non-linearity, imbalance ⇒ $i_N = 0$.

IV-2- Delta (Δ) connection of a three-phase load

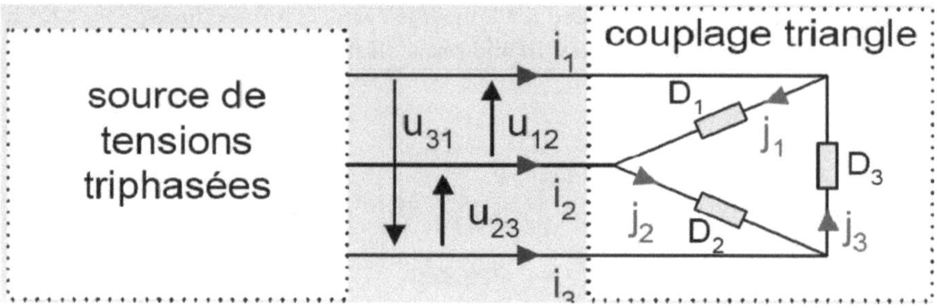

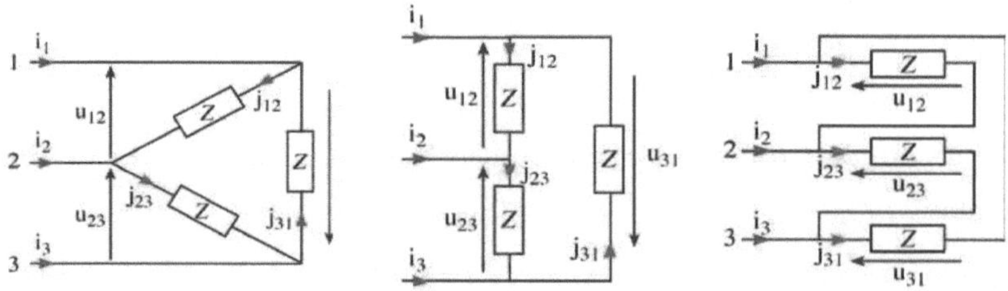

For this connection, there is no neutral.

The currents J1, J2 and J3 are called phase currents.

If the load is linear and balanced, the phase currents form a three-phase current system with RMS values J.

Each phase of the load is subjected to a line-to-line voltage, and I≠J

Therefore, we have :

$$\underline{J}_{12} = \frac{\underline{U}_{12}}{\underline{Z}};\ \underline{J}_{23} = \frac{\underline{U}_{23}}{\underline{Z}}\ and\ \underline{J}_{31} = \frac{\underline{U}_{31}}{\underline{Z}}$$

Thus, in RMS values:

$$\underline{J}_{12} = \underline{J}_{23} = \underline{J}_{31} = \frac{\underline{U}}{\underline{Z}}$$

Therefore, we have :

$$\boxed{I = \sqrt{3}J}$$

- **Fresnel Representation**

$$i_1 = j_1 - j_3$$

$$i_2 = j_2 - j_1$$

$$i_3 = j_3 - j_2$$

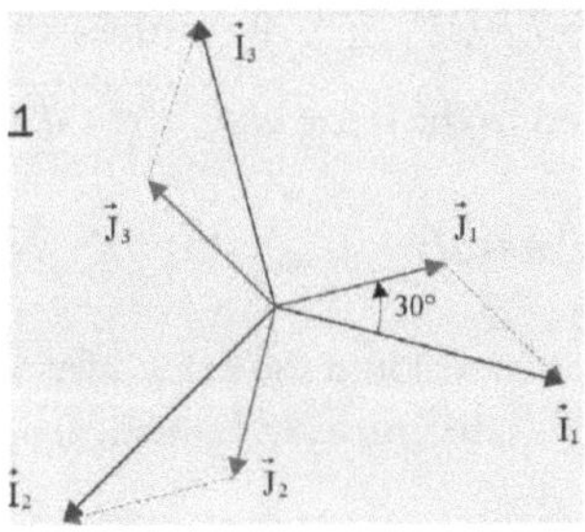

V.Three-phase power

5-1-Reminder in single-phase

- Active power: $\boxed{P = V.I.\ \cos\varphi}$; $[V(V); I(A); P(W)]$
- Reactive power : $\boxed{Q = V.I.\ \sin\varphi}$; $[V(V); I(A); Q(VAR)]$
- Apparent power : $\boxed{S = V.I}$; $[V(V); I(A); S(VA)]$
- Relationship between powers : $\boxed{S^2 = P^2 + Q^2}$

5-2- Power for a star-connected load

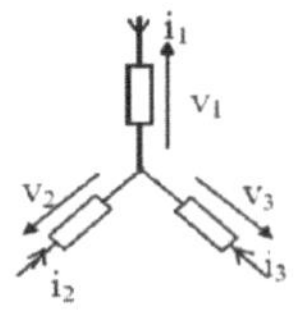

Each phase of the load is traversed by i and has v across its terminals.

So, in the case of a **balanced three-phase load**:

$$P_1 = P_2 = P_3 = V.I.\cos\varphi$$

For the entire load: $P = 3.V.I.\cos\varphi$

Since: $V = \frac{U}{\sqrt{3}} \Rightarrow \boxed{P = \sqrt{3}U\cos\varphi}$

$$Q_1 = Q_2 = Q_3 = V.I.\sin\varphi$$

Similarly: $\Rightarrow Q = 3.V.I.\sin\varphi$

$$\Rightarrow Q = \sqrt{3}.U.I.\sin\varphi$$

Therefore, we have:

$S = \sqrt{3}.U.I = \sqrt{P^2 + Q^2}$ Regardless of the connection type (star or delta), the power is expressed in the same way: $\boxed{P = \sqrt{3}UI\cos\varphi}$; $\boxed{Q = \sqrt{3}.U.I.\sin\varphi}$ and $\boxed{S = \sqrt{3}.U.I}$

VI. Conclusion

This chapter explored the principles of three-phase electrical systems, highlighting their advantages, structure, and practical applications.

Three-phase systems are the standard for electrical power transmission due to their efficiency, cost-effectiveness, and reduced energy losses compared to single-phase systems.

The study of phase and composed voltages demonstrated the balanced nature of three-phase voltages, ensuring stable and continuous power delivery.

Balanced three-phase loads were examined, including star (Y) and delta (Δ) connections, emphasizing their role in maintaining system stability and efficiency.

The analysis of three-phase power provided insights into active, reactive, and apparent power, establishing fundamental relationships for power calculations.

Understanding these principles is essential for students involved in power generation, distribution, and industrial applications, ensuring optimal system performance and reliability.

Magnetic Circuits

Table of contents

I- Introduction

Magnetic circuits are essential in electrical engineering, guiding magnetic flux similarly to how electrical circuits direct current. This chapter introduces magnetic fields, flux, and magnetomotive force (MMF), along with the role of magnetic permeability. It explores linear and nonlinear circuits, highlighting magnetic saturation and hysteresis effects. The study also covers inductance in AC magnetic circuits and its influence on electromagnetic devices. A comparison of hard and soft magnetic materials demonstrates their applications in transformers, motors, and permanent magnets. Understanding these principles is key to optimizing the performance of electrical machines and power systems.

1- Magnetic Fields

Any conductor carrying an electric current is surrounded by a magnetic field. The field (or force) lines follow concentric circles around the conductor.

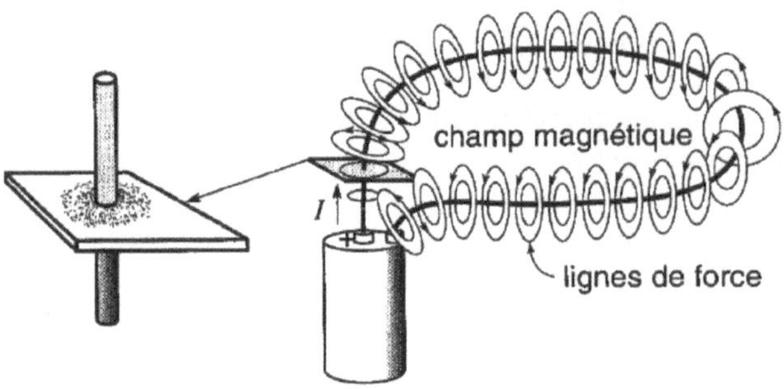

Figure 5.1: Field Lines Around a Conductor

The nature of this magnetic field is revealed by iron filings (Figure 1). The lines of force form circles whose planes are always perpendicular to the direction of the conductor.

The right-hand rule allows for easily determining the direction of the field lines if the direction of the current flowing through the conductor is known (Figure 2).

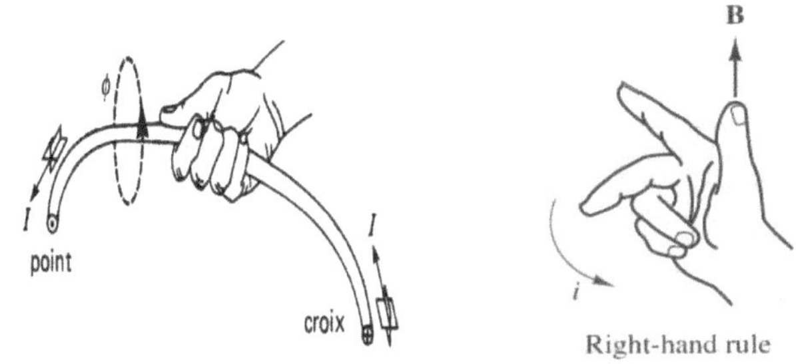

Figure 5.2 Right-hand rule.

2- Flux Density

For a given current, the flux density in Tesla [T] at a specific distance from the center of the conductor is proportional to the current flowing through it. This flux density is independent of the conductor's diameter and material.

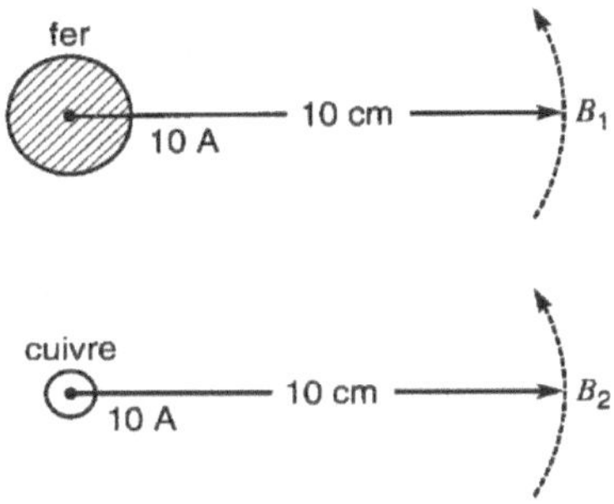

Figure 5.3 Flux density B1 = flux density B2.

The flux density BBB around a straight conductor can be calculated using the formula:

$$B = 2.10^{-7}\frac{I}{d}$$

Where:

B = flux density [T]

I = current [A]

d = distance [m].

Note:

The field created by k conductors carrying equal currents I is the same as the field created by a single conductor carrying a current equal to k**I**.

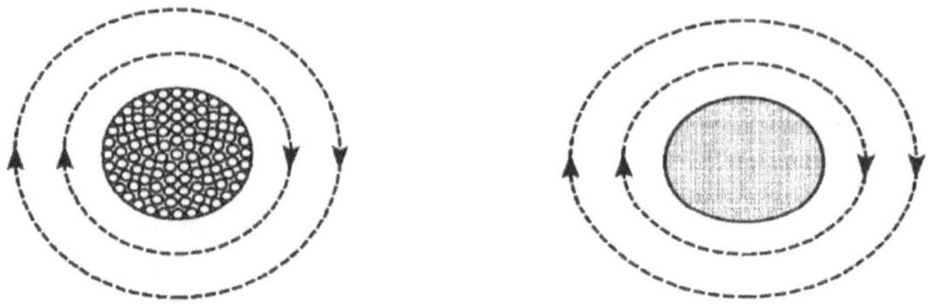

(a) 100 conducteurs de 5 A (b) 1 conducteur de 500 A

Figure 4.4 Equal magnetic fields created by (a) and by (b).

3- Field Produced by a Current in a Loop

Similarly, a coil, and more generally a solenoid carrying a current, is traversed by a magnetic flux whose lines exit through one end, forming the north pole, and enter through the other end, forming the south pole.

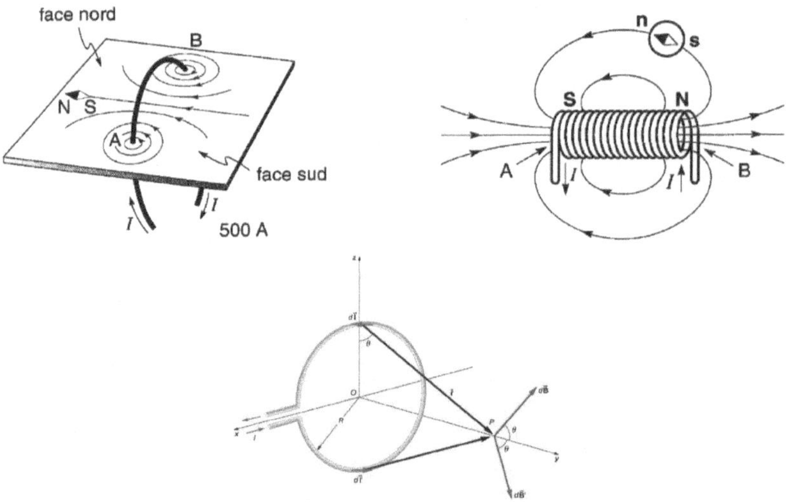

Figure 5.5 Magnetic field created by a current flowing in a coil.

4- Magnetomotive Force (MMF)

The product NI (number of turns × current in amperes) is the magnetomotive force (MMF) developed by the coil.

For example, a coil with 50 turns carrying a current of 8 amperes generates a magnetomotive force (MMF) of:

MMF = 50 turns × 8 amperes = 400 amperes.

A coil carrying a current therefore produces a magnetic field, just like a permanent magnet.

This property is utilized in electromagnets, which consist of a coil wound around a soft iron core.

The high permeability of the soft iron core allows the magnetic field created by the coil to be amplified thousands of times.

Electromagnets have numerous applications in electrical engineering: creating flux in motor and generator poles, circuit breakers, brakes, relays, doorbells, etc.…

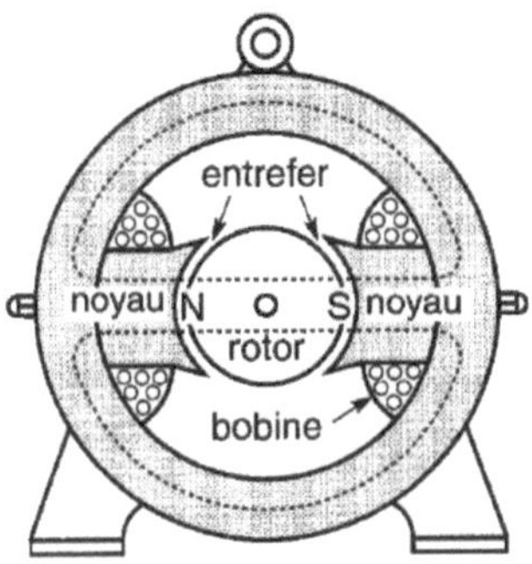

Poles d'une génératrice à courant continu

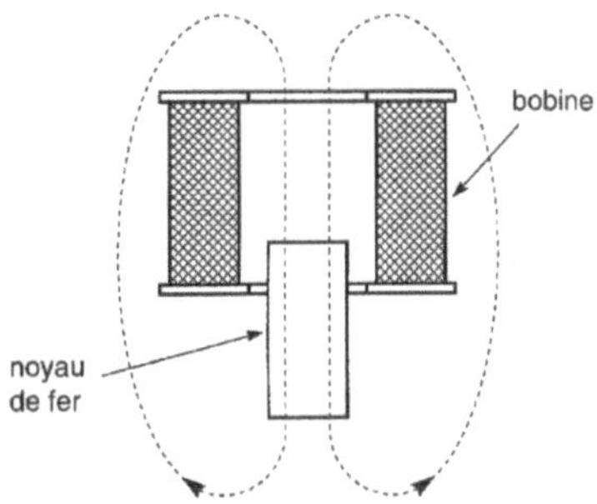

Principe de l'électroplongeur

Figure 5.6 Examples of Electromagnet applications.

II- Magnetic Circuits

Un *circuit électrique* est composé d'un ensemble de d'éléments reliés les uns aux autres dans le but de faire circuler ***un courant électrique***. De la même façon, un *circuit magnétique* est composé d'un ensemble de corps reliés les uns aux autres dans le but de conduire ***un flux magnétique***

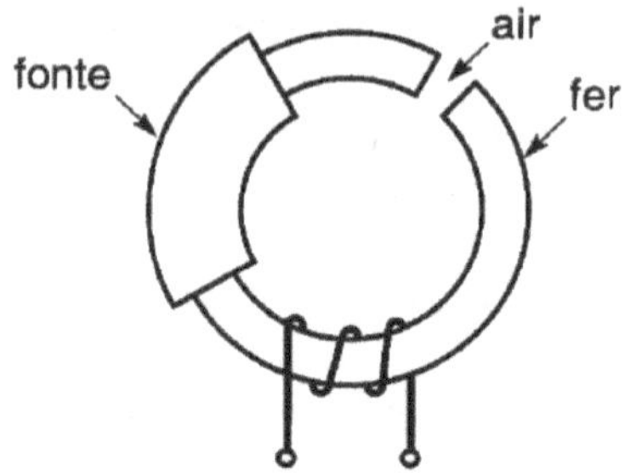

Figure 5.7: Exemple of magnetic circuits

An electrical circuit is composed of a set of elements connected to each other in order to allow an electric current to flow. Similarly, a magnetic circuit is composed of a set of bodies connected to each other in order to conduct magnetic flux.

1- Presentation of Quantities

The table below provides a summary of the quantities involved in linear magnetic circuits, as well as simplified relationships that connect them.

Courant : i	→ Spires et géométrie du circuit	Champ magnétique : H (A / m)	→ Nature du matériau	Induction : B (T)	→ Géométrie du circuit	Flux : ϕ (Wb)
Relations :	**Théorème d'Ampère :** $NI = H \cdot L$	⇢	**Perméabilité magnétique** μ $B = \mu H$	⇢	**Flux** $\phi = B \cdot S$	

2- Magnetic Permeability:

The term μ represents the magnetic permeability of a material. For general materials, this quantity is constant and very low, with a value close to $\mu_0 = 4\pi.10^{-7}(SI)$. (the permeability of free space).

For materials referred to as "magnetic," $\mu = \mu_0.\mu_r$ *où* μ_r is significantly higher. When this quantity is constant, **the material is called linear**.

However, in most cases, the magnetic induction B is limited by a saturation value, referred to as "magnetic saturation," and the value of μ then depends on the value of **H** and, consequently, the current ***i***.

Thus, in most cases: $\mu = \mu(H)$

3- Homogeneous Linear Circuits:

Magnetic circuits are primarily constructed using ferromagnetic or ferrimagnetic materials because they allow for high magnetic flux densities. Indeed, in air or in a non-magnetic material, the field lines produced by a coil carrying a current are not channeled, and the resulting flux is very low. However, in ferromagnetic materials, the field lines are "concentrated" within the material, leading to significantly higher flux values.

The typical configuration of a magnetic circuit is illustrated in Figure 8:

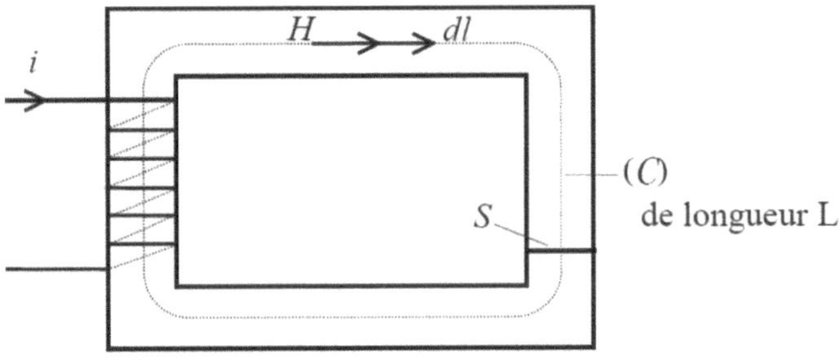

Figure 5.8. Classic Morphology of a Wound Magnetic Circuit.

In this magnetic circuit, since the channeling of the field lines is significant, it is assumed that the magnetic field is constant, particularly along an average curve (represented by a dashed line).
However, Ampère's theorem can be written for this contour:

$$\int_C \vec{H}.\vec{dl} = \int_C H.dl = N.I \Rightarrow H.L = N.I \Rightarrow H = \frac{N.I}{L}$$

We then write, for linear circuits:

$$B = \mu H = \frac{\mu N.I}{L}$$

That is :

$$\phi = B.S = \frac{\mu S.N.I}{L} \quad \Rightarrow N.I = \frac{L}{\mu S}\Phi$$

Now,

$$FMM = NI \Rightarrow FMM = \frac{L}{\mu S}\Phi$$

To establish a practical relationship between the flux and the current that generates it, a quantity called reluctance, denoted by $\Re$, is introduced, which satisfies the so-called ***Hopkinson's law:***

$$\boxed{FMM = NI = \Re\Phi}$$

Pour retenir une relation pratique entre le flux et le courant qui le crée, on fait intervenir la grandeur appelée *Réluctance* et notée satisfaisant à la relation dite ***d'Hopkinson*** :

In summary, to characterize all the quantities in a homogeneous linear magnetic circuit, we will use the following relationship:

$$\boxed{FMM = NI = \Re\Phi \quad \text{Avec } \Re = \frac{L}{\mu S}}$$

4- Analogy with Electrical Circuits

The use of the concept of reluctance allows for an analogy between the relationships of magnetic circuits and those of electrical circuits. The characteristics of this analogy are summarized in the following table.

This analogy will be used freely in linear circuits, making the study of classical magnetic circuits a set of techniques that are easy to master for the electrical engineer.

Circuits électriques	Circuits magnétiques
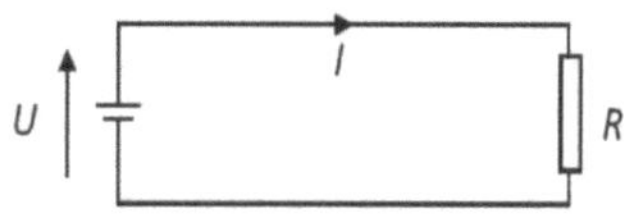	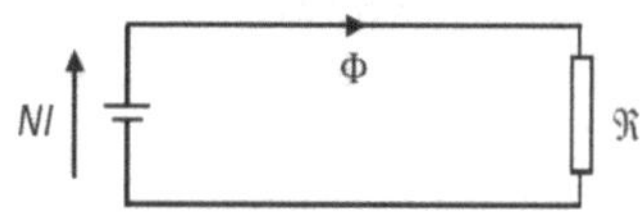
U : Force électromotrice fem	NI : Force magnétomotrice fmm
R : Résistance	$\Re$: Réluctance
Loi d'Ohm : $U = R \cdot I$	Loi d'Hopkinson : $N \cdot I = \Re \cdot \Phi$
Associations de Résistances Série : $R = R_1 + R_2$ Parallèle : $R = R_1 \cdot R_2/(R_1 + R_2)$	Associations de Réluctances Série : $\Re = \Re_1 + \Re_2$ Parallèle : $\Re = \dfrac{\Re_1 \cdot \Re_2}{\Re_1 + \Re_2}$

5- Heterogeneous Linear Circuits

A circuit is said to be heterogeneous when it is made up of different materials or geometries with variable cross-sections. The methodology, as in an electrical circuit, consists of using known reluctance combinations to calculate the different quantities.

Figure 9 shows the cases of **series and parallel heterogeneous circuits**.

For each circuit, the corresponding electrical analogy is also represented, using the symbol $\mathfrak{R}$ to commonly designate reluctance.

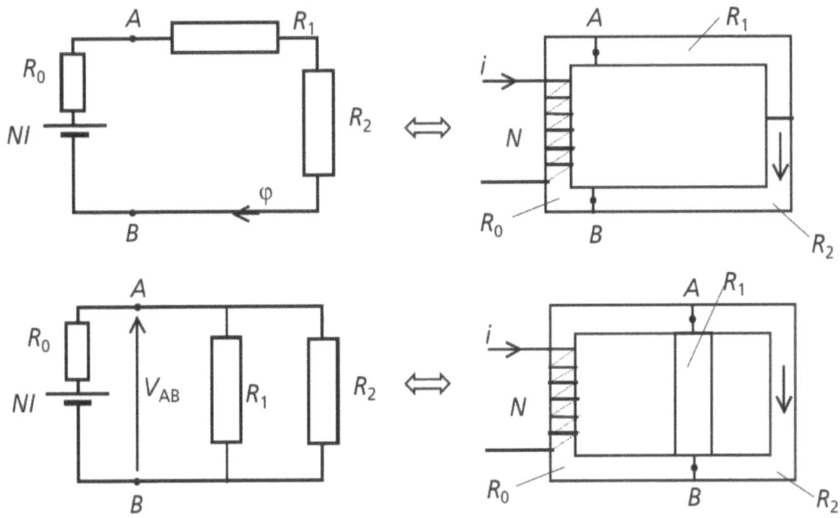

Figure 5.9: Series and Parallel Heterogeneous Magnetic Circuits.

6- Inductance

In a linear regime, inductance is the proportionality constant between the current in the coil and the "total" flux intercepted by the coil, that is, the flux Φ_T.

$$\Phi_T = N.\phi \ .$$

We then write:

$$\Phi_T = N.\frac{NI}{\mathfrak{R}} = L.I$$

The quantity L is the inductance of the wound magnetic circuit, and its unit is the Henry (H).

We retain:

$$L = \frac{N^2}{\Re}$$

7- Non-linear Circuit

When magnetic saturation cannot be neglected in a circuit, it is important to exclude relationships that are only valid in the linear regime.

The only relationships that can always be used are:

- Ampère's theorem: $H.L = N.I$
- The flux / induction relationship: $\Phi = B.S$

However, it is necessary to write: $B = \mu(H) \cdot H$

In practice, for a constant magnetic field ***H***, we refer to the magnetization curve ***B(H)***of the material to match the value of the induction ***B***.

More commonly, we preferentially refer to the curve, which has the same shape as the ***B(H)*** curve, an example of which is presented in Figure 10.

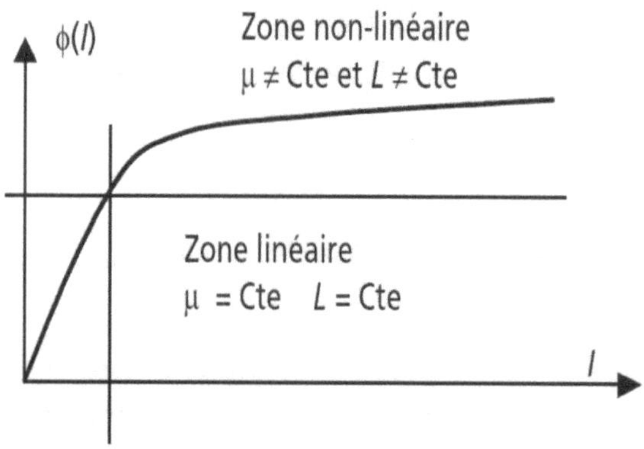

Figure 5.10: Example of Non-linearity of the Curve: Flux/Current.

When the magnetization curve of a magnetic material is known, it is easy to deduce the **relative permeability** μ_r at any given point using the following relationship:

$$B = \mu_0 \mu_r H \Rightarrow \boxed{\mu_r = \frac{B}{\mu_0 H}}$$

Where μ_0 is the permeability of free space, **B** is the magnetic flux density, and **H** is the magnetic field strength.

Figure 11 shows the magnetization curves of three magnetic materials.

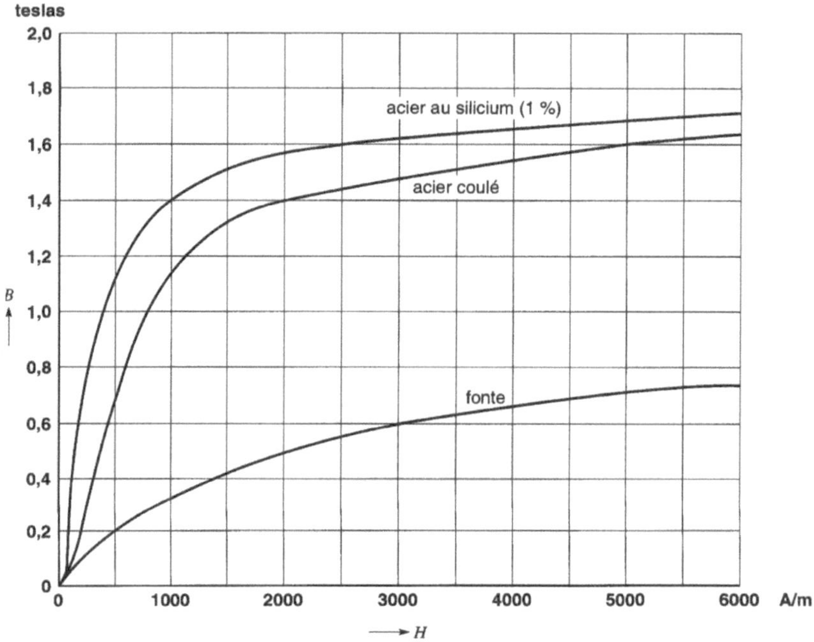

Figure 5.11 Magnetization Curves of Three Magnetic Materials.

8- Magnetic Circuits in Sinusoidal Alternating Current Regime

Consider the circuit below in the sinusoidal alternating current regime:

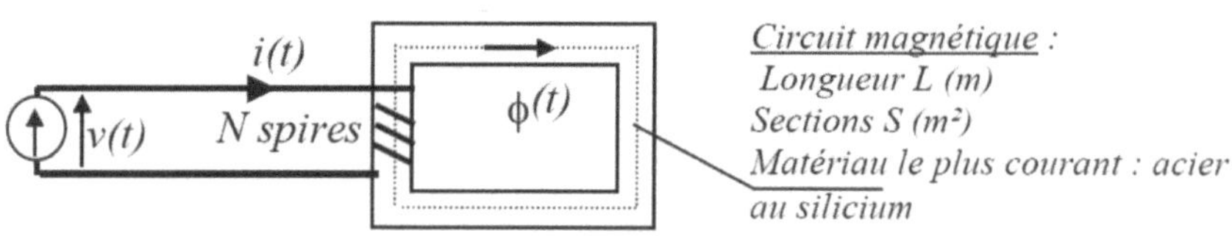

Figure 5.12 Magnetic circuit in sinusoidal A.C regime

With $v(t) = V\sqrt{2}\cos\omega t$

In the sinusoidal alternating current regime, the relationship between the voltage across the coil wound on a magnetic circuit and the flux passing through it is given by Lenz's Law.

A direct relationship appears between the RMS value of the voltage across the coil and the maximum induction (the peak value of the sinusoidal induction). Lenz's Law is written, in the generator convention, as:

$$v(t) = N\frac{d\phi}{dt} = \frac{d(N\phi)}{dt} = \frac{d\Phi_T}{dt}$$

Where **V** is the voltage, **N** is the number of turns in the coil, and **Φ** is the magnetic flux.

Two cases arise:

a- Ideal Linear Material:

If the material has a linear ***B(H)*** curve, this means that the permeability and inductance are constant. From this, we can write:

$$v(t) = \frac{d\Phi_T}{dt} = \frac{d(Li)}{dt} = L\frac{di}{dt}$$

The coil then behaves as a **pure inductance.**

In this case, the relationship between the voltage *V* and the current *I* in the coil follows the formula for inductive reactance in an AC circuit:

$$v(t) = L\frac{di}{dt}$$

Where ***L*** is the inductance of the coil.

- **Relation Voltage/Induction:**

If $v(t) = V\sqrt{2}\cos\omega t = N\frac{d\Phi}{dt}$

$$\Rightarrow \frac{d\Phi}{dt} = \frac{V\sqrt{2}}{N}\cos\omega t \quad \Rightarrow \Phi(t) = \int\frac{V\sqrt{2}}{N}\cos(\omega t)dt = \frac{V\sqrt{2}}{N\omega}\sin(\omega t)$$

On the other hand, $\Phi(t) = \mathrm{B(t).S}$

We then have: $\frac{V\sqrt{2}}{N\omega}\sin(\omega t) = \mathrm{B(t).S}$

That is: $\mathrm{B(t)} = \frac{V\sqrt{2}}{NS\omega}\sin(\omega t) = \frac{V\sqrt{2}}{NS2\pi f}\sin(\omega t)$

We can deduce the maximum value of B(t): $B_{\max} = \dfrac{V\sqrt{2}}{NS2\pi f}$

We end up with: $V = \dfrac{2\pi B_{\max}.NSf}{\sqrt{2}}$;

We will retain **Boucherot's** relation:

$$\boxed{V = 4,44.N.B_{\max}.S.f}$$

b- Nonlinear Real Material:

The real nonlinear material possesses a ***B(H)*** curve that is characterized at low frequency over a cycle of variations and exhibits a hysteresis phenomenon.

This cycle is represented in **Figure 13**.

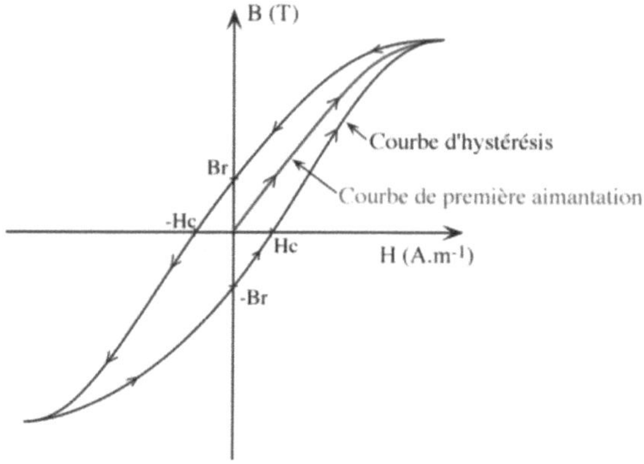

Figure 5.13 : Hysteresis cycle

- **Hysteresis** refers to the lag between changes in the magnetic field strength ***H*** and the resulting magnetic flux density ***B***.

As the magnetic field is cycled, the material's magnetization does not immediately follow the changes in the field but instead retains some memory of past states, which leads to a loop-like curve in the ***B-H*** plot.

- This phenomenon results in energy losses due to the friction-like behavior in the material, represented by the area enclosed by the hysteresis loop.

The **B-H curve** and its hysteresis are key factors when analyzing the performance of magnetic materials, especially in transformers, electric motors, and other electromagnetic devices.

These curves show how a ferromagnetic material responds to the magnetic excitation H.
The initial magnetization curve represents the relationship ***B=f(H)*** when the ferromagnetic material has no prior magnetization.

Hysteresis loop: the curve ***B=f(H)*** when the ferromagnetic material already has magnetization.

- **Linear region:** in this region, ***B=μH*** constant. This is the region typically utilized for transformers and rotating machines.
- **Saturation of the ferromagnetic medium:** when H becomes too large, B hardly changes anymore. The magnetic material is said to be saturated.

While ***B=μH*** still holds, μ is no longer constant.

- **Remanent field Br:** the magnetic field that remains when H = 0 (i_1 = 0).
- **Coercive excitation HcH**: the magnetic field strength ***H*** required to cancel the remanent field ***Br***.
- ***Hysteresis:*** the splitting of the ***B(H)*** characteristic curve of the magnetic material.

*Thus, BBB depends not only on HHH but also on the material's previous magnetization. Ferromagnetic substances have a **"memory"**.*

Note:

Hysteresis originates from the Greek word for "delay" – a delay in demagnetization.

× Hard materials

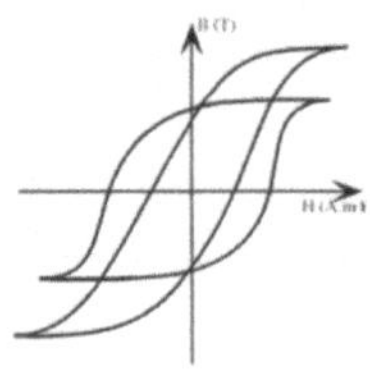

Hard materials, such as steel, are characterized by a strong remanent magnetization that is difficult to reverse, as indicated by a large coercive force (Hc). These materials retain their magnetization even after the external magnetic field is removed, making them suitable for applications where a permanent magnetic field is required, such as in permanent magnets. Due to their high coercivity, hard materials are able to maintain their magnetic properties over time, which is essential in devices like motors, sensors, and magnetic storage systems.

The "wide cycle" refers to the large hysteresis loop of these materials, meaning that the magnetization remains stable over a broader range of external magnetic fields. This property ensures that the material can retain a strong magnetic field without significant loss of energy, making it ideal for creating magnets that need to maintain their strength for long periods. The wide hysteresis loop also indicates that the material requires a substantial external field to demagnetize it, which is a desirable characteristic for permanent magnets used in various mechanical and electrical devices.

Examples of alloys used for magnets.

Ferrite (iron oxide)
Saturation at ≈ 0,6 T
Br ≈ 0,4 T
Hc ≈ 200 kA/m

Samarium-Cobalt (Sm-Co)
saturation at ≈ 1 T
Br = 0,8 T
Hc = 500 kA/m

× Soft materials

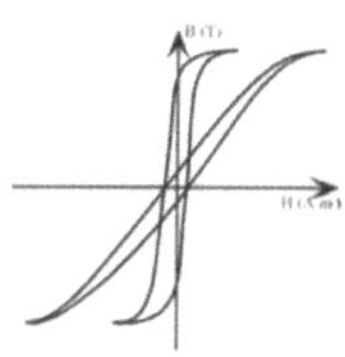

Soft materials, such as iron, exhibit a remanent magnetization that can be easily reversed due to their small coercive force (Hc). This property makes them ideal for applications where rapid magnetization and demagnetization are required, such as in motors and transformer laminations. In motors, they help reduce energy losses by allowing quick changes in magnetization. Similarly, in transformers, their use in thin laminated cores minimizes eddy current losses. The narrow hysteresis loop of these materials means they can efficiently cycle between magnetic states with minimal energy loss, improving performance in electromagnetic applications.

Examples of alloys used for transformer laminations

FeSi 3,5% Si
Saturation at 2 T
Br ≈ 0
Hc ≈ 0
μ_r = 7000 à 50 Hz

Grain-oriented FeSi
saturation at 3 T
Br = 1,4 T
Hc = 8 A/m
μ_r > 40 000 à 50 Hz

Conclusion :

This chapter explored the fundamentals of magnetic circuits, highlighting their importance in electrical and electromagnetic systems.

Magnetic fields and flux density were introduced, along with their calculation methods and practical significance.

The magnetomotive force (MMF) and its role in generating magnetic fields were discussed, emphasizing applications in electromagnets.

The structure and behavior of magnetic circuits were examined, drawing parallels with electrical circuits through concepts such as reluctance.

Nonlinear magnetic behavior, including saturation and hysteresis, was analyzed to understand material properties and energy losses in transformers and motors.

The use of magnetic circuits in AC regimes was explained, detailing inductance and Lenz's Law.

Finally, a comparison between hard and soft magnetic materials highlighted their respective roles in permanent magnets, motors, and transformers.

Single-phase transformers

Table of contents

I- Introduction

On electrical networks, the voltages vary significantly depending on their purpose. The voltages generated by alternators, those used for long-distance transmission, and the voltages required for end-user applications are not the same. The voltage range available within the electrical network is remarkably broad, spanning from as low as 12V for certain specialized applications to as high as 750kV for long-distance transmission lines. This wide range reflects the need to optimize energy transmission efficiency while ensuring compatibility with diverse usage requirements.

Long-distance transmission lines are typically designed as neutral-free systems to reduce costs and minimize losses. However, near consumption areas, it becomes necessary to recreate a neutral point. This is essential to allow the distribution of single-phase lines, which are commonly used for residential and commercial applications. The presence of a neutral enables stable and balanced operation, providing a reference point for the voltage and ensuring safety and reliability.

To address these requirements, a device that combines several key attributes is needed. It must offer high efficiency to minimize energy losses, be cost-effective to ensure economic feasibility, and provide the capability to adjust voltage levels while also recreating a local neutral. The transformer fulfills all these criteria. It is a vital component of modern electrical networks, enabling efficient voltage transformation and ensuring compatibility between high-voltage transmission and low-voltage distribution systems.

II- Ideal Single-Phase Transformer

A single-phase transformer is a fundamental electrical device used to transfer energy between two circuits through electromagnetic induction. It is composed of two windings, known as the primary winding and the secondary winding, which are placed on a shared magnetic core. This magnetic circuit, typically made of laminated iron or ferrite, serves to guide the magnetic flux and improve the transformer's efficiency by minimizing energy losses.

The primary winding is connected to the input voltage source, while the secondary winding delivers the transformed voltage to the load. The number of turns in each winding determines the voltage transformation ratio, allowing the device to either step up or step down the voltage as needed.

An ideal single-phase transformer is a theoretical model used to simplify the analysis of transformer behavior. It assumes perfect operating conditions with no energy losses, which helps in understanding the fundamental principles of operation

Figure 1 provides a simplified schematic representation of a single-phase transformer. It highlights the core structure and the relationship between the primary and secondary windings. This arrangement is critical for ensuring efficient energy transfer while maintaining electrical isolation between the input and output circuits, a key feature that enhances safety and functionality in various applications.

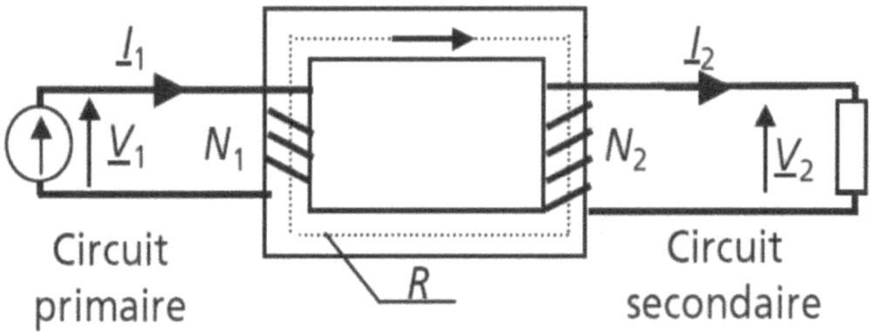

Figure 6.1: Basic schematic of an ideal transformer

The operation of an ideal transformer is governed by two fundamental relationships that describe the behavior of voltages and currents between its windings. The first relationship is the **voltage transformation ratio**, which states that the ratio of the primary voltage to the secondary voltage is equal to the ratio of the number of turns in the primary winding to the secondary winding:

$$\frac{V_1}{V_2} = \frac{N_1}{N_2}$$

The second relationship involves the **current transformation**, which ensures power conservation in the ideal case. It states that the ratio of the primary current to the secondary current is inversely proportional to the turns ratio:

$$\frac{I_1}{I_2} = \frac{N_2}{N_1}$$

These relationships highlight the key principles of electromagnetic induction and energy transfer within an ideal transformer.

$$\begin{cases} v_1(t) = N_1 \dfrac{d\Phi}{dt} \text{ et } v_2(t) = N_2 \dfrac{d\Phi}{dt} \Rightarrow \boxed{\dfrac{v_2(t)}{v_1(t)} = \dfrac{\underline{V}_2}{\underline{V}_1} = \dfrac{N_2}{N_1} = m} \\ et \\ N_1 i_1(t) - N_2 i_2(t) = \Re\Phi \approx 0 \text{ en charge} \Rightarrow N_1 i_1(t) = N_2 i_2(t) \Rightarrow \boxed{\dfrac{i_2(t)}{i_1(t)} = \dfrac{\underline{I}_2}{\underline{I}_1} = \dfrac{N_1}{N_2} = \dfrac{1}{m}} \end{cases}$$

The quantity mmm is called the "transformation ratio

- ## Symboles et conventions :

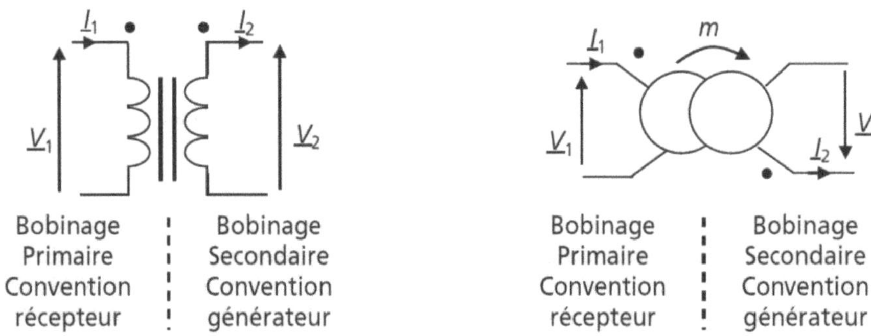

Figure 6.2: Symbols and conventions of the ideal transformer

Figure 2 illustrates the two most commonly used symbols for a single-phase transformer, which are essential for standardizing transformer diagrams in electrical engineering. These symbols follow a specific convention known as the **dot convention,** which helps to define the relative polarities of the primary and secondary windings. The dots next to the windings indicate the polarity of the voltages at the corresponding points of the primary and secondary coils.
The purpose of the dot convention is to provide a reference for determining the phase relationship between the primary and secondary voltages. By convention, the voltage at the dotted terminal of the primary winding is considered to be in phase with the voltage at the dotted terminal of the secondary winding. This helps to ensure that the correct phase difference is maintained when the transformer is used in alternating current (AC) circuits.
Once the polarities are established using the dot convention, the next step is to orient the currents in a way that aligns with the assumed voltage direction. This means that, for the transformer to function correctly in its intended application, the primary winding should always be treated as the receiver (load) and the secondary winding as the generator (source). This ensures that power is transferred properly from the

primary side to the secondary side without causing confusion in phase relationships or current direction.

In this context, maintaining proper orientation of both the voltage and current directions is crucial to avoid errors when analyzing or designing transformer-based circuits, especially in applications involving complex systems like power distribution networks.

It is only by adhering to these conventions that the fundamental relationships apply without any concerns about sign convention. By following the dot convention and correctly orienting the currents and voltages, the mathematical equations that govern the transformer's behavior can be used directly, ensuring consistent and accurate results. If these conventions are not respected, the polarity of the voltages or the direction of the currents may be incorrectly assigned, leading to errors in calculations or in the interpretation of the transformer's performance. This is especially important when analyzing power flow, determining voltage regulation, or performing fault analysis in transformer-based systems. Thus, these conventions serve as a standardized approach that simplifies the design, analysis, and operation of transformers in electrical circuits.

$$\text{In both cases : } \begin{cases} \boxed{\dfrac{v_2(t)}{v_1(t)} = \dfrac{\underline{V_2}}{\underline{V_1}} = \dfrac{N_2}{N_1} = m} \\ \text{and} \\ \text{under load } \boxed{\dfrac{i_2(t)}{i_1(t)} = \dfrac{\underline{I_2}}{\underline{I_1}} = \dfrac{N_1}{N_2} = \dfrac{1}{m}} \end{cases}$$

III- Powers:

The complex **apparent power** at the transformer input refers to the total power that is supplied to the primary side, which includes both the real power (active power) and the reactive power. It is represented by the product of the input voltage and current, taking into account the phase difference between them.

Mathematically, it can be expressed as:

$$\underline{S_1} = \underline{V_1}.\underline{I_1}^*$$

Where $\underline{S_1}$ is the complex apparent power, $\underline{V_1}$ is the primary voltage, and $\underline{I_1}^*$ is the complex conjugate of the primary current. This value is essential for determining the overall power requirements and the transformer's efficiency.

At the output of the transformer, the complex apparent power is equal to the complex apparent power at the input, assuming ideal conditions with no losses.

This means that the total power transferred from the primary to the secondary side is the same, and can be expressed as:

$$\underline{S}_2 = \underline{V}_2 . \underline{I}_2^*$$

Where $\underline{S}_2$ is the complex apparent power at the secondary side, $\underline{V}_2$ is the secondary voltage, and $\underline{I}_2^*$ is the complex conjugate of the secondary current.
In an ideal transformer, the apparent power remains constant across the transformer, and the primary and secondary power are balanced.
In the auther hand:

$$S_2 = V_2 . I_2^* = m.V_1 . \frac{1}{m} . I_1^* = V_1 . I_1^* = S_1$$

Thus, by analogy with the real and imaginary parts, we will note that:

$$\boxed{P_1 = P_2 \text{ et } Q_1 = Q_2}$$

The ideal transformer is completely **passive and lossless**. When it steps up the voltage, it steps down the current (or vice versa), and it does not alter the power that flows through it.
In an ideal transformer, the principle of power conservation holds, meaning the input power is equal to the output power. This implies that any increase in voltage results in a proportional decrease in current, ensuring that the apparent power remains constant. The relationship between voltage and current is determined by the turns ratio, which dictates how the transformer alters these two quantities while maintaining the same total power.
For example, when the transformer **increases the voltage**, it must **decrease the current** by the same factor, maintaining the balance between real power (**P**), reactive power (**Q**), and apparent power (**S**) in the system. As a result, there are no losses in the ideal transformer; the input power is fully transferred to the output, ensuring 100% efficiency. This characteristic simplifies the analysis of electrical circuits, especially in power transmission, as it enables voltage levels to be adjusted without any loss of energy, provided that the transformer is ideal.

Notes :

1- An impedance **Z** in series with the primary of an ideal transformer is equivalent to an impedance $\mathbf{Z.m^2}$ in series with the secondary.
2- An impedance **Z** in series with the secondary of an ideal transformer is equivalent to an impedance $\mathbf{Z/m^2}$ in series with the primary of the transformer.

When we talk about a real transformer, we take into account the losses it introduces as well as its power factor. In this case, it is possible to represent the transformer with an equivalent circuit that includes resistive and reactive components.

In real transformers, losses occur primarily in the form of core losses (hysteresis and eddy currents) and copper losses (resistance in the windings). These losses reduce the efficiency of the transformer, and the power factor reflects the phase difference between the voltage and current, affecting the power delivery.

The equivalent circuit of a real transformer incorporates these losses. The primary winding resistance and the leakage reactance are represented by resistive and inductive elements. Similarly, the core losses are modeled by a parallel resistive element, while the magnetizing reactance is represented by an inductive element. The power factor, which accounts for the phase difference between voltage and current, is included in this circuit to reflect the real-world behavior of the transformer.

This model helps to analyze the actual performance of a transformer under different load conditions and allows engineers to determine efficiency, voltage regulation, and other operational parameters. It also facilitates troubleshooting and optimization of transformer performance in practical applications, ensuring that the power losses and the transformer's response to load changes are properly accounted for.

IV- Real single-phase transformer

The faults and imperfections in real transformers are caused by several factors that influence their performance and efficiency. These faults are represented in the equivalent circuit of a real transformer, starting from the ideal transformer model. The primary factors include:

1. **Series resistances of the windings (R_1 and R_2):** These resistances account for the losses due to the resistance of the primary and secondary windings. These losses are proportional to the square of the current passing through the windings and result in heat generation.

2. **Leakage inductances (L_1 and L_2):** These inductances represent the portion of the magnetic flux that does not link both the primary and secondary windings. Leakage flux results in a reduction in efficiency and contributes to voltage regulation issues.
3. **Core losses (fer losses), represented by R_f:** These losses are due to hysteresis and eddy currents in the magnetic core, which are proportional to the square of the input voltage. These losses cause heat to build up in the core and reduce the overall efficiency of the transformer. The core losses are modeled as a resistance R_f in parallel with the primary winding.

4. **Magnetizing inductance (L_m):** This inductance represents the reactive power required to magnetize the core of the transformer. It is associated with the energy stored in the magnetic field and is necessary for the transformer's operation, even when no load is connected.

The combination of these factors in the equivalent circuit of a real transformer allows for a more accurate representation of its behavior under actual operating conditions. By modeling these losses and reactances, engineers can better predict the transformer's performance, efficiency, and voltage regulation, leading to more effective design and operation of electrical systems.

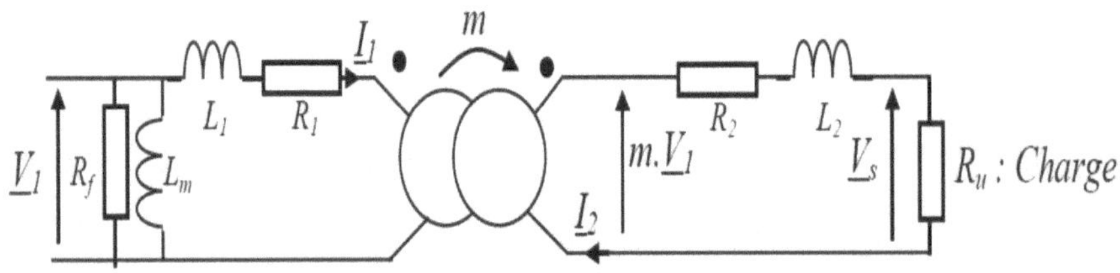

Figure 6.3: Equivalent Circuit of a Real Single-Phase Transformer (Secondary Side)

The complete equivalent circuit of a real transformer includes multiple elements that account for losses and other imperfections, such as series resistances, leakage inductances, and core losses. While this detailed diagram provides a thorough representation of the transformer's behavior, it can be difficult to work with in practical applications. In many cases, especially when quick analysis or basic characterization is needed, this full equivalent circuit is too complex and cumbersome.

However, by making certain assumptions and approximations, such as ignoring minor losses or simplifying the way the reactive components are represented, the equivalent circuit can be simplified for easier use. One common simplification is the "secondary equivalent circuit," where some of the losses and reactances are reflected to the secondary side of the transformer, making the analysis more straightforward. This simplified model retains most of the key characteristics of the real transformer while being more manageable. It allows engineers to quickly assess the transformer's performance, such as voltage regulation, efficiency, and load conditions, without needing to analyze the more complex primary-side components in detail. The secondary equivalent circuit is widely used in practical applications for transformer analysis.

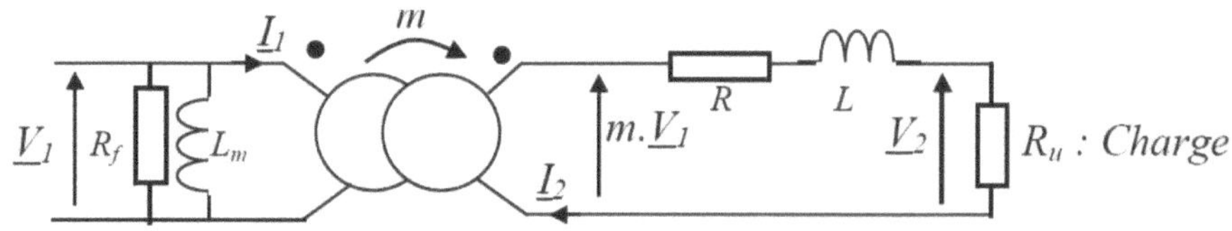

Figure 6.4: Simplified Equivalent Circuit of a Real Single-Phase Transformer (Secondary Side)

Where it is shown that: $R = R_2 + m^2.R_1$ et $L = L_2 + m^2.L_1$

V. Conclusion

This chapter provided an in-depth study of single-phase transformers, focusing on their role in voltage transformation and power distribution.

The ideal transformer was introduced as a lossless model to explain fundamental voltage and current relationships, emphasizing the concept of transformation ratio.

The power equations in transformers were analyzed, demonstrating how apparent, active, and reactive power are transferred efficiently in an ideal case.

The real single-phase transformer was examined, incorporating practical imperfections such as winding resistance, leakage inductance, and core losses.

The equivalent circuit models of real transformers were introduced, allowing engineers to analyze performance, efficiency, and voltage regulation under actual operating conditions.

Understanding transformer behavior is essential for designing efficient power systems, ensuring optimal energy transfer while minimizing losses in practical applications such as power distribution, industrial machinery, and electronic devices.

Introduction to Electrical Machines

Table of Contents

I. General Overview of Electrical Machines

Electrical machines are devices that convert electrical energy into mechanical energy (motors) or mechanical energy into electrical energy (generators). They are essential in numerous fields, including industrial automation, power generation, and household applications. Depending on their function, electrical machines are classified into:

- **Generators**: Convert mechanical energy into electrical energy, commonly used in power plants and renewable energy systems.
- **Motors:** Convert electrical energy into mechanical energy, found in industrial machinery, transportation, and home appliances.
- **Transformers:** Used for voltage conversion, facilitating efficient electrical transmission and distribution.

Electrical machines operate based on **electromagnetic principles**, utilizing magnetic fields to induce motion or generate electrical power. Their efficiency and performance depend on design factors such as the choice of materials, cooling methods, and operating conditions.

II. Operating Principle of Generators and Motors

1. Generators

A generator operates based on **Faraday's Law of Electromagnetic Induction**, which states that a voltage is induced in a conductor moving within a magnetic field. The primary types of electrical generators include:

- **DC Generators**: Common in applications requiring steady direct current, such as battery charging and small-scale power supplies.
- **AC Generators** (Alternators): Widely used in large-scale power generation, supplying electricity to grids and industries.

The fundamental working principle involves:

a) A mechanical force (from a turbine, engine, or another prime mover) rotates a coil within a magnetic field.
b) This movement induces an electromotive force (EMF) in the coil, generating electrical energy.
c) The generated voltage is then delivered to an electrical load for use in various applications.

2. Motors

Electric motors operate on **Lorentz Force**, which states that a current-carrying conductor placed in a magnetic field experiences a force. The main types of electric motors include:

- ✓ **DC Motors**: Known for precise speed control, used in robotics, electric vehicles, and conveyor systems.
- ✓ **AC Motors**: The most common type, further classified into:
  - **Synchronous Motors**: Operate at a fixed speed, ideal for precision applications such as clocks and industrial drives.
  - **Asynchronous (Induction) Motors**: Used in most industrial applications due to their robustness and efficiency.

The working principle of a motor includes:

a. Electrical energy is supplied to the motor windings.
b. A magnetic field is created, interacting with the rotor to generate motion.
c. The generated torque causes rotational movement, driving mechanical loads.

III. Power Balance and Efficiency

1. Power Balance in Electrical Machines

The power balance equation for an electrical machine is given by:

$$P_{input} = P_{useful} + P_{losses}$$

where:

- P_{input}: Power supplied to the machine.
- P_{useful}: Useful mechanical or electrical power output.
- P_{losses}: Power lost due to inefficiencies, which can be classified as:
  - **Copper Losses**: Caused by the resistance of the windings.
  - **Iron Losses**: Due to hysteresis and eddy currents in the magnetic core.
  - **Mechanical Losses**: Resulting from friction and windage in rotating parts.
  - **Stray Load Losses**: Small losses caused by leakage flux and other non-ideal conditions.

IV. Efficiency of Electrical Machines

The efficiency η of an electrical machine is defined as:

$$\eta = \frac{P_{useful}}{P_{input}} \times 100\%$$

Factors affecting efficiency include:

- **Material Quality**: High-conductivity copper and low-loss magnetic materials improve performance.
- **Cooling Systems**: Effective cooling prevents overheating, reducing losses.
- **Design Optimization**: Minimizing leakage flux and improving winding
- configurations enhance efficiency.

Improving efficiency is critical for reducing operational costs and energy consumption, particularly in large-scale applications such as industrial motors and power generators.

V. Conclusion

Electrical machines are fundamental in modern technology, facilitating energy conversion between electrical and mechanical forms. This chapter provided a comprehensive overview of:

- The **types and functions** of electrical machines.
- The **principles of operation** for both generators and motors.
- The **power balance and efficiency** considerations essential for optimizing machine performance.

A solid understanding of these principles enables engineers and technicians to design, select, and maintain electrical machines for various applications, ensuring sustainability, reliability, and cost-effectiveness in power systems and industrial operations.

Bibliographie

1- T. WILDI, ÉLECTROTECHNIQUE 4 ème ÉDITION. Donod.

2- Jacques LESENNE, Francis NOTELET et Guy SEGUIER : *Introduction à l'électrotechnique approfondie*. Technique et Documentation,1981.

3- Pierre MAYÈ : *Moteurs électriques industriels*. Dunod, 2005.

4- MINOT : Cours d'électrotechnique. Polycopié de cours, 1981.

5- Lotfi BAGHLI : Page web de lotfi baghli, 2005. Site internet: http :/ / www.baghli.com. 2005.

6- Jean-Pierre CARON et Jean-Paul HAUTIER : *Modélisation et commande de la machine asynchrone*. Éditions Technip, 1995.

7- James L. Kirtley, Electric Power principles, 2010.

8- John Bird, Electrical and electronic, Elsevier, 2007.

9- Clive Maxfield, Electrical Engineering, Know It, All Newnes, 2008

10 Tahar Neffati, Electricité générale, Ellipses, 2004

11 O. R. Lawrence, Électronique - Principes et applications, McGraw-Hill, 1979

12 Luc Lasne, Exercice et problème électrotechnique, Dunod, 2011

13 Christopher R. Robertson, Fundamental electrical and electronic principles, 2011

14 Mohamed El hawary, Introduction to electrical power systems, Wiley, 2008

15 **Raymond A. Serway, John W. Jewett**, *Physics for Scientists and Engineers*, 6e édition, Brooks Cole.

16 **Max Marty**, *Principes d'électrotechnique*, Dunod, 2005.

17 **Hubert Razik, Abdelaziz Djerdir,** *Électrotechnique et machines électriques***,** Dunod, 2016.

18 **Paul Horowitz, Winfield Hill**, *The Art of Electronics*, Cambridge University Press, 2015.

19 **James W. Nilsson, Susan A. Riedel,** *Electric Circuits*, Pearson, 2018.

20 **Leonard S. Bobrow,** *Fundamentals of Electrical Engineering,* Oxford University Press, 1996.

FSC
www.fsc.org
MIX
Papier | Fördert
gute Waldnutzung
FSC® C083411

Kolibri 360 GmbH
Ferdinand-Jühlke-Straße 7
99095 Erfurt, Deutschland
produktsicherheit@kolibri360.de

Druck:
CPI Druckdienstleistungen GmbH
im Auftrag der
Kolibri 360 GmbH
Ferdinand-Jühlke-Str. 7
99095 Erfurt